Resource Curse Reduction through Innovation - A Blessing for All - The Case of Kuwait

Resource Curse Reduction through Innovation - A Blessing for All - The Case of Kuwait

By

Meshaal Jaber Al Ahmad Al Sabah

CAMBRIDGE
SCHOLARS
PUBLISHING

Resource Curse Reduction through Innovation - A Blessing for All - The Case of Kuwait,
by Meshaal Jaber Al Ahmad Al Sabah

This book first published 2013

Cambridge Scholars Publishing

12 Back Chapman Street, Newcastle upon Tyne, NE6 2XX, UK

British Library Cataloguing in Publication Data
A catalogue record for this book is available from the British Library

ISBN (10): 1-4438-4216-8, ISBN (13): 978-1-4438-4216-7

I dedicate this book in memory of my father
Sheikh Jaber Al Ahmad Al Jaber Al Sabah
who firmly established in me a sense of moral and ethical thinking,
to my mother who gave me the love of life,
to my wife, brothers and sisters who gave me a life of love,
and to my beloved country The State of Kuwait.

TABLE OF CONTENTS

LIST OF TABLES

List of Figures

ACKNOWLEDGEMENTS

A number of people have been influential in the preparation and completion of this book. While the author alone is answerable for the arguments advanced, it would be wrong and ungracious not to welcome this opportunity to thank those people who have given the benefit of their ideas, research, support and attention.

I would like to thank my family, my wife and friends who accorded me moral support and encouragement. I owe special thanks to Dr. Mohammed Al Sabah, to my wife Sara Al Sabah, to my brothers Thamer, Majed, Mishari, Khaled, Khalifa and to my sisters Nouriah and Badria.. I am equally grateful to my friends, Terki, Ali, Abdullah, Hasan, Dr. Al Jedi, Mubarak, Ahmad, Dr. Bilal and Mr. Rageb. I would also like to acknowledge the co-operation of a host of academic specialists, particularly Professor Ralph Paliam and Dr Nabil El-Hilali, as well as several business executives and their secretaries. I am most grateful to those senior government officials and business executives who willingly gave of their time to meet and discuss pertinent issues related to this field. Each of these individuals has shown understanding, support and patience which is truly appreciated.

I would especially like to thank my supervisors Professor Paul Trott, Professor Andy Thorpe and Dr. Salah Al Mazaidi for their scholarly and thorough approach to the subject matter. Their professionalism and patience has played a significant role in the successful realisation of this book. From their enthusiasm and support at the ideas stage to the time spent reviewing or giving constructive criticism and suggestions, they have been a source of intellectual stimulation and academic assistance throughout. This book is a true measure of my thanks for the task my supervisors have imposed on me. It is indeed a privilege to find gentlemen of this stature who combine this with high principles and a real concern for people and it is right that this acknowledgement should rejoice in our achievement.

Last but not least, the patient efforts of Ms. Badia Bougaiz, who had the unenviable task of typing and organizing my notes, are gratefully acknowledged. My sincere appreciation goes to her.

SUMMARY

The strategic issues surrounding the governance of oil resources and its implication for the growth and development of Kuwait through innovation are considered in this study – Resource Curse Reduction through Innovation. Within the large and growing body of work in this area a negative relationship between resource abundance and poor economic performance has often been empirically established. Many of the third world countries are richly endowed with significant natural resources. A plethora of research findings shows that these countries are scoring lower on human development, they exhibit pervasive corruption, display conflicts and a large percentage of their population live in dire poverty. Moreover, an enormous amount of their gross domestic income is spent on defence spending and manifests an autocratic form of governance. For the most part this evidence appears to support the "resource curse" hypothesis. The question that arises is whether there is any prospect of the "resource curse" being converted into a "blessing". This study examines the role of innovation in this context as Kuwait considers moving away from its dependence on its natural resources which sustain the economy. Since, innovation is considered a result of numerous interactions between key organizations and groups in the economy including institutions of learning, government, firms and other organizations which together form an innovation system, it may be opportune to consider the reductive role of innovation related to the resource curse.

There are many unique cultural issues that confront Kuwait, and make it a fundamentally different case from other countries endowed with natural resources. The culture of governance in Gulf countries, and the norms and values within each individual Gulf country, become key determinants of innovation that impact on the various economic, political and social phenomena. By reviewing the extensive literature in both the field of the resource curse and innovation and collecting primary data, this study offers an overview of the challenges of promoting and supporting innovation in Kuwait, and the effectiveness of dissemination of innovative practices throughout the various economic sectors.

Numerous studies have considered whether a country's natural resources are a curse or a blessing. Emerging findings appear to suggest that at times, resource-based economic growth models have indeed inhibited growth rates. Development economics also presents numeric data to substantiate the view that the gifts of nature are non-renewable and cannot be replenished. The hypothesis that natural resources of a country might be more of an economic curse than a blessing needs to be tested at different stages of economic growth of a country. The rate at which natural resources are exploited has often been cause for concern. From an economic perspective, Kuwait should inevitably switch from dependence upon natural resources to the development of sectors based on knowledge, skills, capital and technology. A defining characteristic of many resource-rich countries is the discrepancy between the interest of the stewards of the resources and the owners of the resources. At times those in political office (the stewards) appear to work extremely hard to ensure that the rest of the population (the owners) receive little benefit from the resources with which their countries have been abundantly endowed, and so the governance of natural resources merits further research. The study shows that income accruing as a result of the discovery of oil in Kuwait rapidly changed Kuwait's economic priorities, bringing new opportunities and at the same time new challenges. The findings of the research highlight many important issues relating to innovation and the depletion of non-renewable resources indicating to what extent certain sectors of the economy are innovative. One of the unique challenges facing Kuwait is what collective action is necessary to safeguard time honoured traditions that combine economic prosperity with solidarity. Today Kuwait is in need of new commitments on the part of its citizens and decisive actions in political leadership. Instead of maintaining structures and organizations that have shown themselves unable to deal with the challenges that face Kuwait, Kuwait must be ready to support structural changes. This in particular requires a prioritisation of resources towards education, research and development. Kuwait can only become comprehensively innovative if all sectors support the development of innovative products and services. Strategic issues entailing innovation require the involvement of all parties. These include businesses, the public sector, producers and consumers. A wide-ranging partnership for innovation is necessary, particularly when a country's resources are in question. To establish an optimal framework and develop potential for innovation, the prospect of an innovation-friendly market must be widely accepted and a national innovation system where the flows of technology and information among people, enterprises, and institutions that are the key to the innovation process at the national level are required.

In light of these issues this book recommends the reduction of a resource curse through targeted innovation initiatives. The exploitation of natural assets is a matter of grave concern. Exploration and exploitation are costly and risky exercises in terms of growth and profitability. Kuwait needs to cultivate a culture that fosters creative ideas associated with, among others, safety and security of its natural and human resources, morality, employment and health within the context of an increasingly global environment. A lack of a shared vision, purpose and strategy reduces the vital role that innovation can play. Investment in innovation is therefore critical and Kuwait needs to reinvent itself economically, politically, socially, ethically and morally in this regard. Failure to achieve this would result in Kuwaiti governments failing to fulfill its mandate, and thereby reducing the return to private effort and dampening private initiatives.

CHAPTER ONE

INTRODUCTION AND OVERVIEW OF THE STUDY

"The jar of meal will not be emptied and the jug of oil will not fail"
(*The Prophet's Psalm*)

1.1 Introduction

This book considers the strategic issues surrounding the governance of oil resources and its implication for the growth and development of Kuwait through innovation. Within the large and growing body of empirical work in this area a negative relationship between resource abundance and poor economic performance has often been empirically established. For the most part this evidence appears to support the "resource curse" hypothesis. The question that arises is whether there is any prospect of the "resource curse" being converted into a "blessing". This book places innovation into context within the confines of the natural resource that sustains the Kuwaiti economy. There are many unique issues that confront Kuwait, and make it a fundamentally different case from other countries endowed with natural resources. The culture of governance in Gulf countries, and the norms and values within each individual Gulf country, become key determinants of innovation that impact on the various economic phenomena. By reviewing the extensive literature in both the field of the resource curse and innovation and collecting primary data, this book offers an overview of the challenges of promoting and supporting innovation in Kuwait, and the effectiveness of dissemination of innovative practices throughout the various economic sectors.

1.2 Purpose of this study - innovation in the "blessing-curse" context

Research by Auty and Mikesell (1998) on sustainable development in mineral economies focuses on the determinants of innovation. Several authors, among whom one could include Freeman (1982), Porter (1990), Lundvall (1992) and Nelson (1993) describe the link between innovation and competitive economic outcomes and their results have been widely adopted in the policy research domain. However, the necessary preconditions in formulating and implementing its strategic framework as a curse reductive for oil rich countries has still to be explored, particularly in oil rich Gulf countries. This view is taken on board by Van der Panne and Van Beers (2006) who ask, what favours regional innovation? Economies are inextricably linked to the type of governance and political policies that states have endured over time, with Collier and Hoeffler (2000) eloquently addressing these issues in terms of greed and governance. Wilson (2010:5) in his seminal work – "is it possible to build sustainable innovation capacity in oil rich Gulf Countries" – presents a pessimistic picture and suggests that there is very little scope currently for Gulf States to become more internationally competitive with respect to innovation and knowledge fundamentals.

Any contribution to knowledge must be interdisciplinary, taking into consideration the politics of the region, socio-economic issues, history, and the nature of business conducted on the part of the private and the public sector. Failure to address the issues from an interdisciplinary perspective may result in stakeholders being antagonized and the consequences could be disastrous. Furthermore, an interdisciplinary approach ensures a process of solving broad and complex problems adequately since most significant issues have multiple causes and effects. A range of perspectives should ideally be considered providing a more comprehensive understanding of issues and challenges. The thought that innovation is vital for socio-economic growth and human development was long recognized by Adam Smith and even Karl Marx. As much as the global socio-economic and political environments are in a state of dynamic change, these have a major impact on economies. Economies are in different stages of economic growth. Through oil revenues Kuwait has achieved a status of high mass consumption and at the same time, has faltered in terms of innovation.

Within the context of innovation, countries must nevertheless equate the changes that are taking place in the external environment with time-honoured traditions in order to maintain a certain sense of cohesiveness (Afsaruddin, 2002). Success for any country comes from adopting appropriate changes. Any change that is effectively and consistently managed, presents the country with opportunities for sustained growth and ultimately human and social development. To address the challenges and opportunities presented by today's complex, and often unpredictable markets, an organisation must be able to combine resources in novel ways, dispose of or reconfigure resources that are no longer relevant and acquire new resources. An organisation's ability to manipulate resources continuously and rapidly becomes a competitive capability that is not easily imitated by competitors. This capability to innovate is critical to an organisation's viability since it enables the development and introduction of new products and services. It thus enables an organisation to maintain, or improve its current market position.

Acknowledging that innovation in any economy requires a thorough understanding of its social fabric, geographical environment and political economy, an examination of the potential resultant changes accruing to an economy needs to be considered within a curse-blessing context. Kuwait's blessing, in the form of oil, is placed in the context of the study and can underpin the transformation of Kuwait from a welfare economy to one where free market innovations prevail, resulting in sustained growth and development, with its oil revenues becoming the driver of innovation. The "curses" are in the form of a dependence syndrome that was created as a result of previous forms of governance that pre-date the discovery of oil. Generation upon generation of Kuwaitis have benefitted from the wealth in numerous ways. The lump-sum government hand-outs in the form of donation to the nation, and free medical care abroad and many social and philanthropic gestures by the government are primary causes of concerns since many of these are not sustainable. Innovation initiatives in Kuwait open a new chapter in the conflict over time-honoured traditions and modernity. A haphazard approach in using innovation as a curse reduction tool may have disastrous consequences. Embracing an all-encompassing concept of innovation, this study considers innovation as an idea, as theory and rhetoric, and as a political and socio-economic practice, sustaining an economy and at the same time improving productivity. Moreover, this study places innovation in Kuwait as a policy driven movement, and as a process that shows every sign of reconstituting major institutional sectors of contemporary society. The implementation of innovation becomes a critical component within this definition. In ultra-conservative societies the inertia for change comes from certain policy initiatives created by the government. A mandate by the government to change its *modus operandi* is indeed an impetus for change. An innovative solution by the government in the form of policies and procedures for migration to innovation is the start of change in a mindset towards innovation. In this regard four major institutions have been established by the government to ensure that there is diffusion of innovation in terms of a comprehensive policy and procedure. The institution includes: Kuwait Foundation of Advancement of Science, Kuwait Institute of Scientific Research, Kuwait Investment Authority and the Arab Development Fund. These institutions are considered in greater detail in Chapter Three.

As a result innovation has become a central feature of economic policy in Kuwait and indeed in a variety of economies in the Arab world. Governments worldwide are looking to innovation as a possible solution to their political needs. Innovation in Kuwait should therefore be seen as a process of embracing efficient and competitive initiatives. Effective programmes of innovation in the economy involve many elements, both macro and micro in character. These programmes include the introduction of new goods and new methods of production, the opening of new markets and the sourcing of new material, as well as managing and restructuring enterprises, establishing an appropriate business environment, building financial intermediaries and promoting competitive market conditions. This all-encompassing definition helps to avoid ambiguity and facilitates comprehensive analysis of the issue in all its aspects.

An investment in innovation is an investment in security. The spread of democracy continues throughout the world. Spectacular innovation is affecting the technology of production and the organization of firms as well triggering social and cultural changes. A widely accepted hypothesis is that people are motivated by a hierarchy of needs. Safety and security of the body, morality, employment, health and resources is the second level of human needs identified by Maslow. Satisfaction associated with these issues is said to lead to a sense of belongingness and love and finally to other higher order of needs. An investment in security becomes critical when one considers the broadened view of security in relation to countries endowed with rich natural resources. Since the gifts of nature are not renewable and cannot be replenished, Kuwait needs to align itself to economic, political, social, and ethical strategies that enhance sustainable socio-economic and political development. Insecurity in its broadest definition reduces the return to private effort and destroys private initiatives.

The accompanying Figure 1.1 shows the diffusion of innovation in terms of a wider definition of security that should ideally encompass: the improvement of social relations; the fostering of economic growth, financial development and independence; reducing political, social and financial volatility; fostering employment growth; promoting a sustainable environment; restructuring social security systems; delivering food security, ensuring fiscal discipline; and embracing trade agreements. Education plays a pivotal role in the pursuit of creating this awareness.

Figure 1.1 Diffusion of Innovation – a broadened view

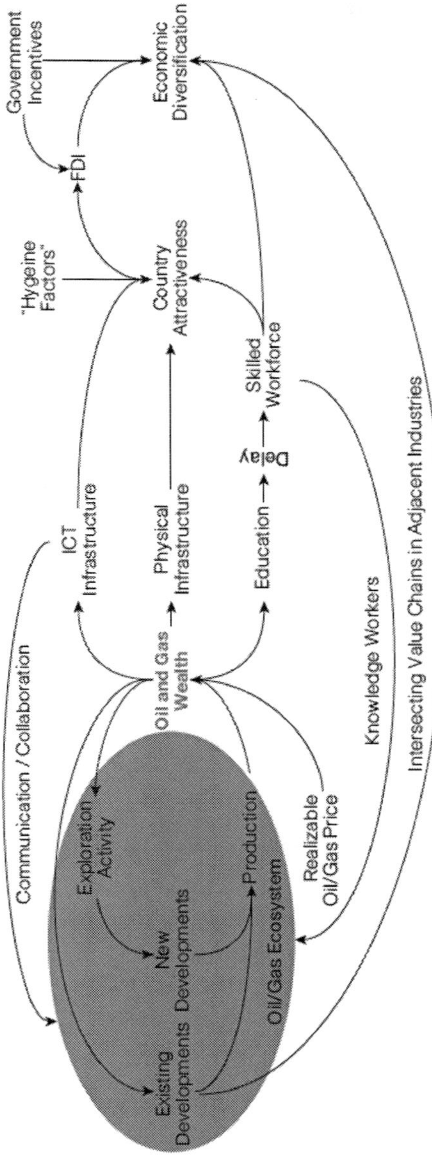

Source: The Natural Wealth of Nations: Transformation of Oil and Gas Producing Economies, Cisco Internet Business Solutions Group (IBSG) (Wood, 2007).

A wider definition of innovation should ideally encompass: the improvement of social relations, the fostering of economic growth, financial development and independence; reducing political, social and financial volatility; fostering employment growth; fostering sustainable environment; restructuring social security; re-focus on food security, ensuring fiscal discipline; and re-energizing trade agreements. An investment in all forms of governance of innovation plays a pivotal role in the pursuit of country attractiveness.

Oil and gas-based economies have an opportunity to use the wealth derived from mineral assets to create societies capable of sustaining growth and diversity, with infrastructure and education taking priority, as illustrated in the figure. Infrastructure will provide the platform for linkages across the oil and gas community, fostering creativity and innovation. Education will help to create an environment where innovation can take place, and will also provide the labor pool so desperately needed by an industry whose specialized knowledge and talent are currently dwindling instead of growing. By creating an ICT infrastructure, educational and research institutions become connected to community and industry. Connected cities become part of the network, providing access to data and enabling collaboration and cooperation among communities, companies, institutions, and government. When the workforce becomes connected, even while mobile or in remote locations, and manufacturing and production are tightly tied to business analysis and decision making, the benefits become tangible and quantifiable.

The purpose of this study is therefore to consider the reductive role of innovation in combating a resource curse situation within Kuwait.

1.3 Defining Innovation within the context of this study

Innovation is one of those words that suddenly seem to be all around us. Firms care about their ability to innovate, on which their future allegedly depends (Christensen and Raynor, 2003), and many management consultants are busy persuading companies about how they can help them improve their innovation performance. Politicians care about innovation too, how to design policies that stimulate innovation has become a hot topic at various levels of government. The European Commission, for instance, has made innovation policy a central element in its attempt to invigorate the European economy (see Chapter 2). A large literature has emerged, particularly in recent years, on various aspect of innovation and many new research units focusing on innovation have been formed (Fagerberg and Verspagen, 2009).

understanding *how* changes in technology contributed to this growth. Schumpeter (1934, 1939, 1942) was among the first economists to emphasise the importance of *new products* as stimuli to economic growth. He argued that the competition posed by new products was far more important than marginal changes in the *prices* of existing products. For example, economies are more likely to experience growth due to the development of products such as new computer software or new pharmaceutical drugs than to reductions in prices of existing products such as telephones or motor cars. Indeed, early observations suggested that economic development does not occur in any regular manner, but seemed to occur in 'bursts' or waves of activity, thereby indicating the important influence of external factors on economic development.

This macro view of innovation as cyclical can be traced back to the mid-nineteenth century. It was Marx who first suggested that innovations could be associated with waves of economic growth. Since then others such as Schumpeter (1934, 1939), Kondratieff (1935/51), and Abernathy and Utterback (1978) have argued the long-wave theory of innovation. Kondratieff was unfortunately imprisoned by Stalin for his views on economic growth theories, because they conflicted with those of Marx. Marx suggested that capitalist economies would eventually decline, whereas Kondratieff argued that they would experience waves of growth and decline. Abernathy and Utterback (1978) contended that at the birth of any industrial sector there is radical product innovation which is then followed by radical innovation in production processes, followed, in turn, by widespread incremental innovation. This view was once popular and seemed to reflect the life cycles of many industries. It has, however, failed to offer any understanding of *how* to achieve innovative success.

After the Second World War economists began to take an even greater interest in the causes of economic growth (Harrod, 1949; Domar, 1946). One of the most important influences on innovation seemed to be industrial research and development. After all, during the war, military research and development (R&D) had produced significant technological advances and innovations, including radar, aerospace and new weapons. A period of rapid growth in expenditure by countries on R&D was to follow, exemplified by US President Kennedy's 1960 speech outlining his vision of getting a man on the moon before the end of the decade. But economists soon found that there was no *direct* correlation between R&D spending and national rates of economic growth. It was clear that the linkages were more complex than first thought.

There was a need to understand *how* science and technology affected the economic system. The neo-classical economics approach had not offered any explanations. A series of studies of innovation were undertaken in the 1950s which concentrated on the internal characteristics of the innovation process within the economy. A feature of these studies was that they adopted a cross-discipline approach, incorporating economics, organisational behaviour and business and management. The studies looked at:

- the generation of new knowledge;
- the application of this knowledge in the development of products and processes;
- the commercial exploitation of these products and services in terms of financial income generation.

In particular, these studies revealed that firms behaved differently (*see* Simon, 1957; Woodward, 1965; Carter and Williams, 1959). This led to the development of a new theoretical framework that attempted to understand how firms managed the above, and why some firms appeared to be more successful than others. Later studies in the 1960s were to confirm these initial findings and uncover significant differences in organisational characteristics (Myers and Marquis, 1969; Burns and Stalker, 1961; Cyert and March, 1963). Hence, the new framework placed more emphasis on the firm and its internal activities than had previously been the case. The firm and how it used its resources was now seen as the key influence on innovation.

Neo-classical economics is a theory of economic growth that explains how savings, investments and growth respond to population growth and technological change. The rate of technological change influences the rate of economic growth, but economic growth does not influence technological change. Rather, technological change is determined by chance. Thus population growth and technological change are exogenous. Also, neo-classical economic theory tends to concentrate on industry or economy-wide performance. It tends to ignore differences among firms in the same line of business. Any differences are assumed to reflect differences in the market environments that the organisations face. That is, differences are not achieved through choice but reflect differences in the situations in which firms operate. In contrast, research within business management and strategy focuses on these differences and the decisions that have led to them. Furthermore, the activities that take place within the firm that enable one firm seemingly to perform better than another, given the same economic and market conditions, has been the focus of much research effort since the 1960s.

The Schumpeterian view sees firms as different – it is the way a firm manages its resources over time and develops capabilities that influences its innovation performance. The varying emphasis placed by different disciplines on explaining how innovation occurs is brought together in the framework in Figure 1.1. This overview of the innovation process includes an economic perspective, a business management strategy perspective and organisational behaviour which attempts to look at the internal activities. It also recognises that firms form relationships with other firms and trade, compete and cooperate with each other. It further recognises that the activities of individuals within the firm also affect the process of innovation. While there are many arguments and debates in virtually all fields of management, it seems that this is particularly the case in innovation management. Very often these centre on semantics. This is especially so when innovation is viewed as a single event. When viewed as a *process*, however, the differences are less substantive.

The collection of quantitative data of a study of this nature immediately requires an explicit definition of fundamental terms. Innovation is one such term in this study and "resource curse" is another. Bakken (2002) identifies innovation as a fuzzy concept that evokes sharp political reactions. According to Schumpeter "radical" innovations shape big changes in the world, whereas "incremental" innovations fill in the process of change continuously. The term innovation certainly covers a vast range of ideas and policies relating to change. This may include privatization which Star (1988:1) considers may vary from the "eminently reasonable to the wildly impractical". Varied, and at times unclear in its meaning, innovation has unambiguous political origins and objectives too.Innovation goes far beyond research and development. The impact of innovation extends beyond the confines of research laboratories to users, suppliers and consumers. Governments, private businesses, non-profit organizations and other institutions are beneficiaries of innovation or innovators themselves. According to the Oslo Manual of 2005 which is the foremost international source of guidelines for the collection and use of data on innovation activities across the OECD countries (Organisation for Economic Co-operation and Development) contend that the ability to determine the extent of innovation initiatives and the characteristics of innovators are prerequisites for the pursuit and analysis of policies aimed at fostering innovation. The Manual investigates the field of non-technological innovation and the linkages between different innovation types. Furthermore, it includes an annexure on the implementation of innovation surveys in developing countries.Innovation is at the heart of economic change. Schumpeter proposed a list of various types of innovations. These include: introduction of a new product or a qualitative change in an existing product; process innovation new to an industry; the opening of a new market; development of new sources of supply for raw materials or other inputs; and changes in industrial organisation.

The first issue related to innovation is product innovation. This involves a good or service that is new or significantly improved which may include significant improvements in technical specifications, components and materials, software in the product, user friendliness or other functional characteristics. The second refers to process innovation and relates to new or significantly improved production or delivery methods. Marketing innovation, the third form of innovation includes significant changes in product design or packaging, product placement, product promotion or pricing and finally, organisational innovation relates to a new organisational method in business practices, workplace organisation or external relations.

The characteristics of innovation suggest that innovation is the result of numerous interactions between key organizations and groups in the economy including universities, government, firms and other institutions, which together form an innovation system (Wilson, 2010). Wilson (2010) goes on to add that innovation does not take place within a vacuum and that there is an interaction between numerous stakeholders. According to Yam et. al. (2011) a national innovation system consists of flows and relations which exist among industry, government and educational institutions in the development of science and technology. Moreover, they consider Technological innovation as a learning process that results in enhancement of the knowledge and skills firms need to choose, install, operate, maintain, adapt, improve, and develop their technology requirements. Yam et. al. (2011) add that in a world of increasing competition and technological change, the generation and diffusion of innovations increasingly rely on new technological knowledge generated not only through internal research and development department, but also by the firm's interaction with external sources of innovation, particularly in the region in which the firm operates.

1.4 Innovation and the resource curse hypothesis

It is widely acknowledged that revenues resulting from natural resources should generate wealth for an economy, promote economic progress, and increase the wellbeing of each citizen. The logic behind this assertion can be based upon simple common sense, which suggests that while money cannot buy happiness, it is a good "down payment". A large windfall of revenues accruing to an economy from an abundance of natural resources ought to place that economy in pole position economically compared with others. While not a central concern of this study, the question as to whether resource-rich countries have experienced a worse performance (in terms of economic progress and poverty reduction) than countries without such apparent "benefits" is an important issue. However, this study concentrates on understanding innovation as a strategy that converts the blessings into a common good for all. Much of this conversion process relies upon configuring governance structures that make innovation possible. The "Resource Curse" phenomenon is not an immutable law, but the study considers it a strong recurrent tendency.

Berkhout et al. (2006), in their seminal work, identify the changes that are taking place in the so-called *innovation economy*, in which - besides capital, labour and knowledge - creativity is identified as the fourth principal factor of production. The authors go on to describe the activities in an innovation economy as creative enterprise with knowledge. Creativity is an important aspect of human endeavour, particularly in distressing economic times. The assumption of the study is that together with creativity, innovation creates added value. This may be central to converting curse effects into blessings. Whether innovation is an imperative for the survival of economies endowed with natural resources and whether it is a solution to the political, social and economic woes of Kuwait are considered in this study.

For much of the contemporary period the Kuwaiti economic system entailed the provision of free housing and other welfare services to Kuwaiti nationals. Security in the form of jobs in the public sector is a privilege that is accorded to Kuwaitis and the private sector is not in a position to offer the same salaries. This has created a dependence syndrome resulting in a "financial duty" and the state's obligation to fulfill this dependence syndrome. This is becoming untenable as a result of the increasing population and the fact that Kuwait's fiscal revenues remain limited to the revenues generated by oil. Between 1990 and 2009 the Arab Planning Institute shows that the population increased by 2.44%. Many of the stateless community are being incorporated in the State of Kuwait. In this regard the Arab Planning Institute suggests that the continued welfare system will become unsustainable. Oil price fluctuations make government revenues unstable and therefore this study considers whether innovation can help combat the adverse effects of this violability.

On a no-change scenario, the future of this country will be uncertain, possibly bleak. The population is set to more than double by 2035. Four-fifths of Kuwaiti citizens are employed in the public sector. Oil reserves are large, but will decline over time. Kuwait has now reached a stage where its entrepreneurs feel so constrained that they look abroad, and not at home, for opportunities, (Rt. Hon. Tony Blair, 2010).

It is widely believed that Innovation serves as an effective economic mechanism for achieving greater efficiency, strengthening the role of the private sector, improving the public sector financial health, and freeing up resources for allocation to other important areas of government. Kuwait, like many countries around the world, has already taken these views into account and started revitalizing its public enterprises.

1.5 Systematic approach

The process of resource-curse reduction (or resource-blessing enhancement) through innovation requires a methodological approach. Key decision makers must base their decisions on both their goals associated with curse reduction and the respective socioeconomic and institutional environment which determines the strategies adopted. In this regard, it was essential that individual interviews or focus group discussions with key stakeholders be held.

Individual interviews or focus group discussion is a research technique that collects data through interaction with the interviewer on a specific topic. These interviews were particularly useful as empirical evidence and resources relating to curse-blessing effects in Kuwait are limited. In the study, the individual interview method was firstly used to gain an in-depth understanding of key stakeholders' perceptions relating to Kuwait's oil production and its implications for innovation. Secondly, the interview identified the challenges that face the economy. This assisted in formulating the hypothesis for the study, and helped develop a questionnaire that reflects the sentiments of key stakeholders from the private and public sectors.

The findings of the preliminary study provided insights on:

- The extent to which resource abundance (in the form of oil) shaped the Kuwaiti Economy;
- Whether Kuwait uses its natural resources in effective ways;
- Perceived weaknesses of an oil dependent State;
- Options to reduce oil dependency in Kuwait;
- Thoughts of stakeholders regarding the Kuwait Investment Authority's policies and roles;
- Whether privatization should be an option;
- Kuwait's potential for innovation; and

- Factors associated with a well-diversified and technologically advanced economy.

1.6 Chapter outline

This initial chapter presents an overview of the study, placing it into perspective and introducing its purpose. Since innovation is considered a key determinant in the reduction of curse-effects, the chapter identifies the objectives of the study and briefly introduces the methodological approach used.

Chapter Two is a comprehensive literature survey which focuses on the resource curse. This chapter considers the findings of recent studies which have examined the issue of whether a country's natural resources are a curse or a blessing and considers the emerging findings that appear to suggest that, at times, resource abundance has indeed inhibited growth rates. The hypothesis that natural resources of a country might be more of an economic curse than a blessing is a cause for concern. From an economic perspective, economies endowed with rich natural resources should inevitably switch from dependence upon natural resources to the development of sectors based on knowledge, skills, capital and technology. The chapter also considers whether a boom in resource revenues can lead to an enduring competitiveness.

Chapter Three places Kuwait in context within a political and economic framework and considers the institutional framework expected to convert valuable natural resources into enhanced standards of living for Kuwaiti citizens. The unique issues that confront Kuwait, and make it a fundamentally different case from other countries endowed with natural resources are addressed in this chapter. The culture of governance, and the norms and values within Kuwait become key determinants of innovation. A review of the institutional environment thus offers an overview of the challenges of promoting and supporting innovation in Kuwait.

Chapter Four, surveys resource curse reduction through innovation. Income accruing as a result of the discovery of oil in Kuwait rapidly changed Kuwait's economic priorities, bringing new opportunities and at the same time new challenges. The chapter highlights that the government of Kuwait is concerned about many important issues relating to the depletion of non-renewable resources. It identifies the unique challenges facing Kuwait and considers what collective action is necessary to safeguard time honoured traditions that combine economic prosperity with solidarity. This chapter proposes a framework that is favourable to innovation. The prospect of an innovation-friendly market is also considered.

The final and concluding chapter of the study, presents the recommendations and also identifies the associated limitations of the research. Moreover, the challenges that were encountered are discussed and possible areas for future research are also provided. The chapter concludes by proposing there is a critical need for a planned investment in innovation - so as to convert the natural resources of Kuwait into a greater blessing for all.

1.7 Conclusion

The purpose of this book is to consider the role of innovation in combating a resource curse situation within Kuwait. The chapter recognizes that innovation has become a central feature of economic policy in a variety of economies in the developed and developing world, with governments around the globe looking to innovation as a possible solution to their political and economic goals. Innovation in Kuwait should therefore be seen as an ongoing process of embracing efficient and competitive initiatives. A compelling aspect of the study is change through innovation, with innovation considered the engine of development, as no economy can continue to evolve and thrive without a strong innovation framework.

CHAPTER TWO

LITERATURE SURVEY: THE RESOURCE CURSE

"It's a blessing and a curse.
Watch out, one don't make things worse.
For the best things used ill can become evil;
And the worst things should prove to sell, if used well".
(Palliam, 2011)

2.1 Introduction

This chapter considers the findings of recent studies which have examined the issue of whether a country's natural resources are a curse or a blessing. The resource curse is a term which generally refers to the negative growth and development outcomes that can result from the development of natural resources. Emerging findings appear to suggest that at times, resource-based economic growth models have indeed inhibited growth rates. Development economics also presents numeric data to substantiate the view that the gifts of nature are non-renewable and cannot be replenished. The hypothesis that natural resources of a country might be more of an economic curse than a blessing needs to be tested at different stages of economic growth of a country. The rate at which natural resources are exploited is often a cause for concern. From an economic perspective, economies endowed with rich natural resources should inevitably switch from dependence upon natural resources to the development of sectors based on knowledge, skills, capital and technology.

A hugely valuable resource, oil is a black, viscous substance and a motor of global industrialization. For Kuwait in particular, it offers the potential for exceptional profit generation benefiting both the state and private sectors. As a result oil is responsible for what has often been called the "paradox of plenty". This chapter looks at this issue whilst considering the ability or inability of richly endowed countries to raise their growth rates and development levels above those of countries without natural resources. Moreover, the chapter considers whether a boom in

Innovation is the management of all the activities involved in the process of idea generation, technology development, manufacturing and marketing of a new or improved product. There is a distinction between an innovation and a product. Product is an output of innovation. This is consistent with what Drucker (1992) posits: ".... In a knowledge economy knowledge is a product, in a knowledge-based economy, knowledge is a tool". Today, the idea of innovation is widely accepted. It has become part of our culture – so much so that it verges on becoming a cliché. But even though the term is now embedded in our language, to what extent do we fully understand the concept? Moreover, to what extent is this understanding shared? A scientist's view of innovation may be very different from that of an accountant in the same organisation.

A brief analysis of economic history, especially in the United Kingdom, will show that industrial technological innovation has led to substantial economic benefits for the innovating *company* and the innovating *country*. Indeed, the industrial revolution of the nineteenth century was fuelled by technological innovations (*see* Table 1.1).

Table 1.1 Nineteenth-century economic development fuelled by technological innovation

Innovation	**Innovator**	**Date**
Steam engine	James Watt	1770–80
Iron boat	Isambard Kingdom Brunel	1820–45
Locomotive	George Stephenson	1829
Electromagnetic induction dynamo	Michael Faraday	1830–40
Electric-light bulb	Thomas Edison and Joseph Swan	1879–90

Technological innovations have also been an important component in the progress of human societies. Anyone who has visited the towns of Bath, Leamington and Colchester will be very aware of how the Romans contributed to the advancement of human societies. The introduction over 2,000 years ago of sewers, roads and elementary heating systems is credited to these early invaders of Britain.

Innovation has long been argued to be the engine of growth. It is important to note that it can also provide growth almost regardless of the condition of the larger economy. Innovation has been a topic for discussion and debate for hundreds of years. Nineteenth-century economic historians observed that the acceleration in economic growth was the result of technological progress. However, little effort was directed towards resource revenues leads to an enduring competitiveness. It also highlights that with continued dependence of an economy upon natural resources governments should diversify the economy and secure other revenue sources. Countries and regions with an abundance of natural resources, specifically point-source non-renewable resources like minerals and fuels, tend to have less economic growth and worse development outcomes than countries with fewer natural resources. This is hypothesized to happen for many different reasons, including a decline in the competitiveness of other economic sectors (caused by appreciation of the real exchange rate as resource revenues enter an economy), volatility of revenues from the natural resource sector due to exposure to global commodity market swings, government mismanagement of resources, or weak, ineffectual, unstable or corrupt institutions (possibly due to the easily diverted actual or anticipated revenue stream from extractive activities). Finally, the extent to which the resource curse impacts on the Kuwaiti economy is addressed.

2.2 Conceptualizing the resource curse and blessing

Generally gifts of nature are defined as natural resources. These range from *inter* alia, mineral deposits, water, arable land and vegetation, to natural forests, marine resources, animal life and oil. By definition natural resources are fixed in supply and therefore inelastic. These gifts are therefore non-renewable or exhaustible assets. In the case of oil, most oil reserves lie in developing economies where governance mechanisms tend to be weak. This predicates a high prevalence of the resource curse syndrome. According to Smith (1776):

Projects of mining, instead of replacing the capital employed in them, together with the ordinary profits of stock, commonly absorb both capital and stock. They are the projects, therefore, to which of all others a prudent law-giver, who desired to increase the capital of his nation, would least choose to give any extraordinary encouragement. (Smith, 1776, p.562).

The "paradox of plenty" relates to any country that is richly endowed in natural resources and at the same time presents itself with numerous conflicts. These conflicts could range from social and political disorders to economic woes of the majority of people. As a result an abundance of natural resources presents curse effects and encourages conflict. Nevertheless there are studies (Ross, 2003) that suggest oil-producing developing countries, that have very highlevels of oil revenue are remarkably stable. An analysis of the ways in which governments spend oil revenues gives an indication of the extent of the paradox.

Auty (1998, p.9), reflects the view of numerous authors arguing that resource-rich and developing countries have not performed as well as countries that are deficient in resources. Nabli and Silva-Jauregui at the Fourteenth World Congress of the International Economic Association in Morocco (Marrakech) held between August 29 and September 2 2005. analyze the relationship between democracy or its deficit in the MENA (Middle East and North Africa) region. They argue that there is no question that democracy has lagged behind in the MENA region. There is a persistent democracy gap. Contrary to what can be observed in the rest of the world, there is no correlation in the MENA region between the level of income and progress in democracy. The rich oil exporting countries in particular have among the lowest democracy scores. Per capita income growth in the MENA region has also been low, though not as low as Sub-Saharan Africa. However, the general literature on the link between democracy and growth is not very conclusive and many of its results are either fragile or conditional. While there has been little democratic progress in MENA countries there has been progress in human development, in particular in education and health. They discuss a battery of governance indicators and conditions for good governance under democracy. Countries like Iraq, Lebanon and Syria that are more fragmented have worse initial conditions for good governance. Oil producing countries usually have worse governance indicators. The authors are skeptical that good governance may come out of non-democratic regimes. On the other hand, obstacles to reform are numerous and political economy factors would tend to favor the status quo. This supports the idea of a resource-rich environment having an adverse effect on the growth of GDP. While this inverse correlation between growth and resource abundance has received widespread acceptance, recent research from Gupta (2007) and Karnick and Fernandes (2009) suggests a positive association between growth and resource abundance. However with the findings of Sachs and Warner (2001) and Auty (2001) showing strong inverse relationships between resource abundance and economic growth, this is the principal premise of the resource curse hypothesis.

Models could be generated to determine whether countries are better or worse off with smaller or larger endowments of natural resources (and whether any econometric associations are subject to bias). However, the character of the resources themselves is more important to consider than the underlying reasons for any association. Should countries fail to build upon their resource base productively and exert caution in its use, the result would be a failure in development that can be attributable to a country's overt dependence on resources. Moreover, Cowen (2007) points out:

> ... it is unfortunate that economists have to debate whether natural resources are a blessing or a curse for a developing nation. Minerals, diamonds or oil may appear to represent automatic wealth but resource-rich countries usually become mired in corruption. High oil revenues, for instance, allow a government to maintain power and reward political supporters without doing much for its people. (Cowen, 2007).

Yet Wright and Czelusta's (2002) findings suggest that some nations with large extractive industries – like Malaysia, Botswana and Chile – have overcome the resource curse and introduced sound development strategies that address poverty alleviation. Norway has also been cited as a country that has surmounted "the curse", along with Indonesia (prior to 1997), Australia and Canada. Auty (1998, p.46) questions if these exceptions exist, can it then be true that "the problems of mineral economies are inherent to the production function of mining?"

When one considers and addresses how a development strategy exploits a country's abundance of resources, the argument is often reduced to a discussion of politics. Arguably countries whose political and social institutions and structures have not succeeded in supporting sustained development tend to be those who suffer from civil discontent, corruption and conflict. One can well imagine that in an environment of fragile institutions and factional politics, resource abundance may be a mixed blessing. The problem however lies not with the resources themselves, but in how the resources are managed. Whilst there may be strong evidence of the resource curse syndrome, there do however remain some compelling examples of economies who have converted the curse into a blessing. Resource curse presents itself in numerous different ways. These include political conflicts, social anomalies, and economic iniquities as a result of wealth accruing from natural resources. The poor governance of natural resource could be the major cause of conflicts. In this regard Kuwait in particular faces numerous challenges and these are considered in the subsequent sections.

2.3 The presence of a resource curse

Successfully managed resource economies surveyed in this chapter show that countries can overcome any "curse" bestowed upon them. Given the existence of a negative relationship between poor economic performance and resource abundance, countries rich in resources are usually measured against whether they have overcome this curse. Van der Ploeg (2011) writing in the Journal of Economic Literature considers whether Natural Resources are a curse or blessing. Their empirical evidence suggests that either outcome is possible. The paper investigates a variety of hypotheses and supporting evidence for why some countries benefit and others lose from the presence of natural resources. Van der Ploeg (2011:366) contends that a "resource bonanza induces appreciation of the real exchange rate, deindustrialization, and bad growth prospects, and that these adverse effects are more severe in volatile countries with bad institutions and lack of rule of law, corruption, presidential democracies, and underdeveloped financial systems". Another hypothesis is that a resource boom reinforces rent grabbing and civil conflict especially if institutions are bad, induces corruption especially in nondemocratic countries, and keeps in place bad policies. Finally, resource rich developing economies seem unable to successfully convert their depleting exhaustible resources into other productive assets. Van der Ploeg (2011) also offers some welfare-based fiscal rules for harnessing resource windfalls in developed and developing economies.

However, the resource curse hypothesis is highly sensitive to the chosen time period, the nature of the natural resources and the methodology used in arriving at such conclusions. As poor economic performance impacts on poverty and can also trigger conflicts and political change; abundant natural resources, rather than propelling growth and development, have often undermined it by promoting poor governance, conflict and pervasive poverty. The situation in many countries provides ample evidence for the dispiriting curse.

The empirical literature refers to countries that are endowed in rich natural resources, including various forms of fuel and minerals. The findings in various studies(Harford and Klein, 2005; Mehlum, Moene and Torvik, 2002; Karl, 1997) show negative relationships between poor GDP performance and an abundance of natural resources. Sachs and Warner in their numerous publications over the years have applied different techniques to try and isolate the effects of a number of potential explanatory variables. Whilst researching ninety five developing countries over a 20 year period starting in 1970, in 1997 they found a negative correlation between growth and natural based exports, and they remain convinced that resource abundance does have some deleterious effects on economic performance.

Similarly, Auty, (2001a, p.3) found that " … between 1960 and 1990, the per capita incomes of resource poor countries grew between two to three times faster than those of the resource abundant countries". While the difference between rich mineral driven countries and the weakest performers is greater than was expected, such findings are sensitive to the time-period chosen. Prior to the 1970s quantitative data shows that resource-rich countries grew more than countries that lacked natural resources (Stevens, 2003). Sharp fluctuations in the price of minerals can also distort the results. In Auty's 2001 study, for example, oil prices fell from $42.70 in 1999 to $20.04 in 2000 and the per capita GDP captured this fall.

Discussions on how to define 'resource abundance' also raises problems. Definitions include: dependence on primary products, labour force employed in the sector and population size. Export orientation has also been used to define resource abundance. Limi (2006) identifies Botswana as a resource-rich diamond exporting country, whose growth rates over several years have been remarkably strong – 7.8 percent since the 1980s (forty percent of which can be attributed to mining). Other empirical work by De Ferranti et al. (2002) suggests that the negative growth effect is a result of the resource production dependence creating export dependence. In general empirical findings suggest that poor per capita growth performance was manifest in mineral exporting countries. In addition to a negative growth impact mineral rich countries have a very poor record in eradicating poverty and other social evils. This supports the commonly held belief that an abundance of resource has a negative impact on income inequality (Auty, 1994b; Fields, 1989; Sarraf and Jiwanji, 2001).

Carneiro (2006), citing reports of the United Nations Development Programme (UNDP), suggests that resource-abundance frequently leads to increased income inequality. Two reasons are given as to why this might occur. Firstly, oil, gas and mining industries are often characterized by their 'enclave' nature, with few forward and backward linkages into the economy. During production, such industries employ only a relatively small number of highly-skilled, well-paid workers (and generally import the majority of inputs). Arguably more damaging is the possibility that what opportunity there is for manual and semiskilled jobs (most specifically during construction) may be lost through the influx of cheap foreign construction labour and the trend towards global procurement and sourcing. Secondly, in such circumstances public expenditure may exacerbate inequality. This can occur where expenditure is concentrated in the formal sector in towns and cities, skewing distribution against rural households (or where it is orientated towards the interests of the wealthier classes, for example favouring the construction of a university over investment in rural roads). As a consequence of these factors society identifies the production and export of natural resources with the interests of the rich. Carneiro (2006) also argues that there is a relationship between the abundance of oil, gas and minerals and limited success in poverty alleviation. Examples include worsening infant and child mortality and life expectancy at birth. The effects are possibly more pronounced for non-fuel mineral producers.

A further dimension of the resource curse is that it engenders conflict in societies as identified by Ross (2001) and Collier and Hoeffler (2000) leading to reduced growth rates and growing poverty. One can advance several reasons to explain this. Firstly, a society may view resource rents as something worth fighting for and to plunder (depending on who is in power). Indeed the revenues generated encourage those in power to increase spending on the military to counteract potential threats. Revenue generation can also alienate the population, particularly where separatist movements exist. These separatists include different religious and social groups. Environmentalists who view the environmental damage that occurs as a result of resource exploitation can also constitute a major separatist group. Not all literature supports this view however. In research covering one hundred and nine countries between 1957 and 1990, Smith (2001) argues that revenues generated from oil extraction do not necessarily lead to the infighting that is predicted.

Secondly, a conflict between those in power and those governed leads to internal civil wars and strife, which impacts adversely on poverty. The poor are affected more than the rich, as the latter have the resources to surmount the crisis, while the former lack coping strategies. Thus the hardships of the poor are exacerbated by wars and conflict which absorb important resources that could otherwise be fruitfully used in improving socio-economic performance and hence to alleviate poverty.

Thirdly, natural resource abundance can strongly mitigate political change as stated by Auty (2001b, p.10-11) and can help underpin autocratic regimes. Consequently, an abundance of resources can pervert institutions that were democratic and transparent and suppress dissent so that power is not strongly contested. Opaqueness in public finances and corruption may well become widespread.

Fourthly, the resource curse must be considered in terms of its regional or local impact, and this aspect has tended to be neglected in the literature. The environmental damage caused to regional and local communities can be devastating and, while some funds may flow from national governments to local and tribal authorities, very few resources generally filter down to those most afflicted.

While the prevalence of a curse is noted in many countries, Stevens (2003) also identifies countries with large oil, gas and mineral revenues that have avoided such a curse, but who may be considered vulnerable. This is due to their dependency upon the receipt of large oil and mineral revenues. If such revenues fall sharply the negative impact of this will threaten the financial viability of the countries themselves. Stevens (2003) then links this vulnerability to infant mortality, life expectancy and illiteracy via a physical quality of life index. His results support the hypothesis that some countries avoided the curse, and led him to suggest that strategies could be formulated to overcome any resource curse effects.

Why might countries suffer from a resource curse? Stevens (2003) identifies six potential transmission mechanisms:

i long-term declines in trade;
ii revenue volatility;
iii Dutch disease;
iv crowding out effects;
v increasing the role of the state;
vi and finally the socio-cultural and political impacts.

Logically, natural resources should ideally promote economic growth and development, since natural capital expands the production possibilities of an economy. The next section examines the relationship between natural resource abundance and the resource curse in terms of the six transmission mechanisms identified by Steven (2003). Clearly while the whole issue of what determines whether resource abundance is a curse rather than a blessing is complex, the six transmission mechanisms provide some understanding into the resource curse hypothesis.

2.4 Resource curse: the transmission mechanism

The previous sections presented a detailed discussion relating to a paradox of plenty: large revenues accruing from natural resource exploitation seem to have brought little or no prosperity to the countries that are laden with natural resources. This is consistent with the findings of several authors including Karl (1997) and Cowen (2007). Instead natural resource prevalence has caused poverty, conflict and socio-economic decline. Human rights abuses and environmental deterioration may be important issues, but the main economic problem is an influx of revenue. As a result real exchange rates rise and the country's industrial and agricultural export products become uncompetitive, a phenomenon commonly referred to as

'Dutch Disease'. In this regard the extent of declining terms of trade needs to be considered, together with revenue volatility and crowding out effects.

According to Frankel (2010) oil wealth need not necessarily lead to inferior or superior economic or political developments, through any of these channels. Rather, it is best to view oil abundance as a double-edged sword, with both benefits and dangers. It can be used for ill as easily as for good. That mineral wealth does not in itself confer good economic performance is a striking enough phenomena, but the priority for any country (and Kuwait in particular) should be in identifying ways to avoid the pitfalls that have afflicted other resource rich economies in the past.

2.4.1 Declining Terms of Trade

Raul Prebisch (1950) and Hans Singer (1950) in separate studies analyzed movements in the price of primary products and manufactured goods over time using data on prices, wages and manufacturing mark-ups. The results show evidence of a negative effect on the terms of trade of primary producers, which more than offset the positive effects arising due to increasing volume of primary production output. Their findings provided an explanation for the negative net trends in the terms of trade of primary producers. Nevertheless, a slow decline in the terms of trade does not necessarily explain the sort of deterioration in economic performance which has conventionally been associated with the resource curse. Both Singer (1950) and Prebisch (1950), for example, propose that technological progress and economic growth in the rich countries increased the demand for manufactured products relative to the demand for agricultural produce, as consumers with higher incomes could only spend a limited amount of their increased purchasing power on more food. This, for them, may be a more important explanation behind declining terms of trade than the resource curse.In the In the Kuwaiti case, Salisu and Yagudin (2007) found that the shortage of other (non-oil) raw materials in fact resulted in an increase in trade opportunities. Although oil spurred the first industries in Kuwait, they contend that oil did not generate much in the way of new industries locally. Instead, they argue, shortages of these other raw materials have created opportunities for FDI investment in Kuwait based on vertical integration, although they acknowledge too that industrial development in the country faces formidable bureaucratic obstacles. Moreover, although oil revenues have been volatile over time (see next sub-section), this has been more problematic for oil exporters than a continual decay in the terms of trade.

2.4.2 Revenue volatility

Several authors, including Auty (1998) and Mikesell (1997), suggest revenue volatility as an explanation for the resource curse. These theories rely upon the fact that oil, gas and mineral revenues have been extremely volatile over relatively short periods of time. During the years 1972-1992, for example, Mikesell (1997) found that regions with high primary export experienced much greater trade volatility than industrial countries.

Several challenges arise as a result of trade volatility. It is extremely difficult to pursue a prudent fiscal policy. The result is investor uncertainty and spending policies that are erratic. In this regard Sachs and Warner (1997) found little evidence to support the idea that resource rich countries have higher savings rates over time to counter this. Gylfason et al. (1999) noted that there is an inverse relationship between the levels of domestic investment and primary product exports, suggesting that any unexpected gains are more likely to be consumed than saved/invested. It could therefore be suggested that reductions in economic activity in an economy may well be of benefit - in so far as it encourages economic reforms that would not be likely to occur in less stressful conditions (Stevens, 2003).In the case of Kuwait in particular, oil has come to dominate life and therefore dependence upon a commodity whose price is highly volatile creates understandable concern. Chapter Three shows that the most dramatic element of Kuwait's economic development has been the steady and rapid expansion of its oil industry since the 1970s. By the mid-1980s Kuwait was refining four-fifths of its oil domestically and marketing some 250,000 barrels a day. By the 1980s oil production and gave Kuwait one of the highest per capita incomes in the world.

Garaibeh (1987) suggests, however, that revenue volatility has not to date impacted negatively upon growth and development and the poverty reduction initiatives of the Kuwaiti government. This is attributable to the country's strong sovereign wealth fund which has been created from oil revenues. According to many estimates, Kuwait's fund is now worth approximately $300 billion, a sum that can take care of the pernicious effects of revenue volatility.

2.4.3 Dutch Disease

The term 'Dutch Disease' was introduced by economists who found a link between the discovery of large deposits of natural gas in Holland and a decline in the Dutch manufacturing sector. As a result of the resources boom, the guilder appreciated (due to increased gas export sales), but this adversely affected other (non-gas) exporters. The resource boom also attracted scarce capital and labour inputs to gas production and away from other sectors and, as a result, the output of other traded goods fell. In a seminal paper Corden and Neary (1982) considered the impact of an oil boom and identified two effects, the spending effect and the resource movement effect. In the case of the latter, a higher marginal product in the booming resource sector causes factors of production to move into the oil sector, pushing up the rewards for these factors and causing other sectors to contract. The spending effect is the result of the high demand in both tradable and non-tradable goods sectors generated by the increased resource rents. Contraction of the tradable (non-oil) sector is inevitable. One governmental response is to use subsidies to protect the non-resource tradable sector, but this aggravates the sector's problems by making it subsidy-dependent.

Al-Sabah (1988) reflecting upon the Dutch Disease in Kuwait, argues that, although the Dutch Disease was experienced in Netherlands in the 1960s following gas discoveries, the diseases is not however exclusive to oil-resourced countries only, but rather it has also occurred in other non-oil exporting countries. For instance, Japan's manufacturing sector's technological advance boom in the 1960s adversely affected less dynamic tradable sectors, including the agriculture sector. Also the booming of Swiss bonds and money exports in the 1970s resulted in real appreciation of the Swiss franc hurting the country's traditional exports and export-competing industries.

The characteristics of the economy in question determine in which direction production and productivity in both tradable and non-tradable sectors can go.[1]Although potentially exposed to Dutch Disease, with oil revenues driving the exchange rate upwards, oil price volatility has in fact resulted in a series of local booms and busts (notwithstanding the 'sterilising' activities of the country's sovereign wealth fund) as prices and supplies of the commodity have fluctuated a great deal in world markets. As a consequence, the country is more exposed to exchange rate uncertainty - that can be harmful to exports and other trade (including foreign investment) – than a continually strengthening exchange rate as characterized the Dutch experience.

[1] Auty and Evia (2001) suggest it was highly significant in Bolivia, to the point of very little production taking place in non-tradable sectors. They found that the two classic symptoms of Dutch Disease held, namely an appreciation of the real exchange rate and a crowding out of the non-oil traded goods sector.

2.4.4 Crowding out effects

Related to Dutch Disease, 'crowding out' essentially refers to any reduction in private consumption or investment that occurs because of an increase in government spending. Should the increase in government spending be financed by a tax increase, there would be a tendency to reduce private consumption. Should there be an increase in government spending, a reduction in private investment may occur. There are, however, conflicting opinions on how the private sector reacts to more government spending (Auty, 2001b). Central to the dispute is how well financial markets transfer funds from those who have (and do not wish to spend) to those who do not have (and wish to spend). There is also controversy on how financial markets react to more government borrowing (Carneiro, 2006). This issue is especially sensitive in those countries that are in transition, where the level of government intervention in the economy is reduced – in terms of whether there is indeed a commensurate increase in private consumption.

In Kuwait in particular, the government began to play a major role in all industrial and socio-economic development. The first major push for state-led industrialization occurred with the establishment of the Ash Shuaybah Industrial Zone in 1964 which comprised electricity and water distillation plants, expanded port facilities, metal works, and plants manufacturing chlorine, asphalt, cement, pilings, and prefabricated housing. The government also provided the necessary infrastructural facilities such as roads, gas, electricity, water, sewerage, port facilities, communications, and rented or leased industrial sites to private entrepreneurs at nominal rates.

According to the 2011 IMF report Kuwait's economy has grown steadily as a result of government spending. According to the report, the economy is expected to grow over the medium term as the government implements the development plan and the global recovery supports demand for oil. However, its $133 billion economy is dominated by the state oil industry, the biggest employer outside the public service. Moreover, as Kuwait's vast stock of foreign assets and the return of these assets insulate the economy from any short-term shocks, such as a sudden oil price drop, strong state control in the economy and heavy dependence on the oil sector are unlikely to decrease in the short to medium-term.

2.4.5 Role of the State

Where oil, gas and minerals are the property of the state, the resource revenues in the first instance accrue to the government - who are then mandated by the citizens to ensure that the revenues are used appropriately. Some argue government intervention is particularly necessary in rich-resource laden countries. When the government uses these rents to finance increased expenditure (or to cut taxes), private sector investment is crowded out by way of higher interest rates. State spending is always a controversial issue, not only in terms of how much should be spent, but also in terms of the details of its distribution and funding. As the public sector often forms a large part of the economy, such government spending has major implications on the micro and macro political economy. Spending invariably also impacts on the quality of the living conditions in the economy. Therefore, while the role of the state is to raise the overall growth performance of the economy, and to increase the chance of its citizens to escape from poverty, there is also concern as to the extent of state intervention in the economy. State intervention can strengthen human capabilities and reduce transaction costs, but it can also dampen innovation and entrepreneurial activity. The critical challenge is how to strike the right balance between state and private economic involvement and also between state spending that focuses primarily on growth and spending that aims at sustained poverty reduction.

Currently, Iraq, the neighbouring country of Kuwait, is determining what form Iraq's government should take. One crucial question is how to handle its vast oil wealth. Oil riches in Iraq are far from the blessing perceived before the liberation by the US. Oil growth is not only seen to be negatively correlated to democracy in Iraq, but oil revenue is viewed as impeding the development of institutions and values critical to open, market-based economies and political freedom, civil liberties, the rule of law, the protection of property rights and political participation.

The government of Kuwait is the ultimate custodian of the oil wealth of the country. Kuwait's oil wealth has run parallel with development of their political and economic structures, facilitating numerous social services provided by the state to the citizens of the country.

2.4.6 The socio-cultural and political impacts

Oil-related peace and tranquility (and also violence on the other hand) have been topics of discussion in both a domestic and international context vis-à-vis the resource curse. With regard to the key contextual conditions responsible for peace and violence in oil rich states, research underlines the basic relevance of cultural cleavages and political-institutional and socio-economic weaknesses that existed even before the discovery of oil. Oil has indirectly boosted the patriotism and, in some cases, heightened the risk of violent conflicts. Scholarship has proposed a resource curse hypothesis, arguing that an abundance of natural resources stimulates dysfunctional economic policy choices and rent-seeking, creates conflicts over the distribution of wealth, and can ultimately result in civil war (Cowen, 2007). Kuwait's attempt to circumvent the resource curse through state institutions that bring oil wealth under political control guarantees a large income to the state from oil production. Such strong government control over the production of oil, and the subsequent establishment of an oil fund invested abroad, have strengthened citizens' expectations of benefits they may receive. Yet, as much of the wealth is locked up in the oil fund and is not immediately available, political trust becomes an issue of concern. Auty and Gelb (2001) have developed a typology of states based on whether they are homogenous or factional (comprising several ethnic groups), as well as benevolent or predatory. Kuwait is a benevolent state since it tends to maximize social welfare, invest in infrastructure and human capital. In the Auty-Gelb (2001) typology a benevolent state has an accountable executive; an efficient civil service; adheres to the rule of law; allows the participation of 'civil society' in policy making; and has an open and transparent policy making process. Predatory states in contrast promote rent-seeking, lobbying and uncompetitive industrialization.

2.5 Decision making within the confines of the resource curse

Generally it is considered that a large windfall in revenues leads to poor decision-making by government (Auty, 2001b). The prevalence of natural resources in an economy raises expectations among the citizens of the economy. The citizens exert enormous pressures on a government to make decisions and, as a result, hasty and uncoordinated decisions may often be made. Spending distortions affect the way the economy functions (Auty, 2001b). Moreover, an abundance of resources tends to reduce government prudence and procedures for due diligence are ignored. Haphazard decisions become manifested in government decision making, and leads to governments spending without due thought. Furthermore, decisions on spending tend to be concentrated in fewer hands and may encourage unacceptable ethical conduct, including corruption and rent seeking on the part of government decision makers.

2.5.1 Management of resources for the common good

The difficulty of adequately managing oil revenues is compounded by several factors. Of these the most important is the lack of transparent political institutions necessary for counteracting corruption or rent-seeking. According to Carneiro (2006) the key elements of democratic governance should ideally include executives who are accountable, an efficient civil service and tax authorities, independent legal systems, active and informed civil societies, and open and transparent policy-making processes. Within such an environment business transactions are transparent and trade continues to take place without restrictions. Should such practices fail, earnings are squandered, precious assets are depleted and widespread poverty prevails. Moreover, fundamental socio-economic problems can arise out of a widespread pursuit of self-interest. Resulting socio-economic problems require resource to the notion of the 'common good'. This notion originated in the writings of Plato, Aristotle, and Cicero, where 'common good' is considered as governance that is to everyone's advantage in equal measure. The common good in oil-rich countries consists primarily of having systems of management that benefit all people. However, on occasion, time honoured traditional Arab values of 'common good' – which may not be consistent with current Western best practice - may be more proactive in furthering innovation, development and growth.

Oil, gas and mineral projects are usually associated with the involvement of foreign companies in developing economies. Undoubtedly corruption in trade is a great cause of annoyance and misfortune to foreign companies and, if good governance is the hallmark of progress, corruption is its bane. Serious concerns surrounding governance standards in oil rich countries have been raised (Auty-Gelb, 2001). The credibility of government officials is placed in doubt as stakeholders question their ethical and moral obligations. This can be avoided if governments ensure that there are checks and balances in place. Engeli and Pieth (2000), in a paper commissioned by the World Bank, present research carried out in the transitional states, and conclude that bribery often pays when used either to secure large-scale procurement contracts, or to buy influence. The existence of powerful economic incentives, makes bribery all the more difficult to address and underlines the need for rigorous enforcement of anti-bribery legislation. Concerns with corruption, over and above the moral and ethical discourse, include:

- Impact on development and the poor: bribery and corruption inhibit development as it is the poor who largely pay for the cost of bribes, either through higher prices, or lower quality services. Bribery also creates a democratic deficit as key decisions affecting citizens are made away from the public arena for reasons outside the public interest.
- Impact on markets: bribery and corruption distorts competition.
- Integrity of public services: both nationalisation and liberalisation policies increase the opportunities and incentives for bribery and corruption, thereby undermining public confidence in the integrity of public services.
- Impact on workers and whistleblowers: whistle-blowers provide a mechanism for detecting bribery and thus potentially provide a powerful tool for deterring corruption. Whistleblowers need to be properly protected both by national legislation, and at a corporate level, through the establishment of appropriate disclosure channels.

While corruption is widespread in many countries, it is not clear whether corruption is more likely in resource rich countries. Legend has it that an absence of corruption in some oil rich countries would result in their sovereignty ceasing to exist. Rent-seeking behaviour also imposes significant losses on many economies and may impact negatively on economic reforms and impede innovation. Why should rent seeking be greater in countries with large oil, gas or mineral revenues? It appears the greater the funds available in the public purse, the less noticeable the leakages to stakeholders. Auty (1998) argues that rent seeking is more prevalent in oil rich countries, because wealth is concentrated in a politically dominant public sector (or possibly in a small number of companies).

However, corruption and rent seeking can benefit economies - should the benefits be used for productive investment within the economy. In this instance the outcome will be very different than if they are used for the furthering of one's own selfish interests.

2.5.2 The characteristic of investment decisions in oil rich states

Effective investment decisions in oil-rich countries may be hampered by mismanagement at all levels. While part of this is attributable to poor governance structures, investment decisions can, at times, strongly support the growth and development of an unproductive economic base. Sarraf and Jiwanji, (2001) have outlined that poor investment decisions are evident in many countries. Poor investment decisions include greater investment in non-tradable sectors, like investment in the military, indulging in prestige projects ranging from constructing palatial residences and airport infrastructure which are quite inappropriate for the economy. As large oil revenues allow governments the luxury to borrow on the strength of these revenues, this can also militate against sound decision-making.

2.5.3 Industrialisation and innovation policy

A government policy for industrialisation and innovation should be directed towards fulfilling certain goals and objectives.

These may include the following:

- What path should industrial development take?
- How can sustained growth in productivity be maintained?
- How can gainful employment, particularly for citizens, be expanded and optimal utilisation of human resources achieved?
- How can social and economic disparities be reduced?
- How can poverty be eradicated and self-reliance attained?
- How can international competitiveness be attained?

Policies of oil-rich states such as Kuwait must ideally reflect the realisation of these goals. A policy of industrialisation and innovation ought to answer these concerns, and should ideally propose initiatives towards employing revenues for national reconstruction and development. Corden and Neary (1982) raised the concern that in the case of numerous resource oil-rich countries, they have generally failed to promote competitive manufacturing sectors. However many economists, including Wagenast (2007), Ross (2006) and Auty (2001), have since considered competitive manufacturing a primary source of technological progress and innovation, and this certainly has major implications for economic development and growth.

When an economy adopts an industrial policy based on greater state intervention for the import substituting sector, this can help oil-rich countries break out of the cycle of underdevelopment - and develop markets that are powerful and efficient while serving the public interest.

2.6 Curse effects – political and social issues

A commonly held view, and a prevalent thread throughout the reviewed literature, appears to suggest that countries which exhibit extreme dependence on natural resources, such as oil, are always vulnerable to various forms of conflict and civil war (Ross, 2006). However, the impact of natural resources on social capital and institutional structures also needs to be addressed. Ross (2006) suggests that resource rich countries accumulate social capital at a far slower rate than poor countries. An explanation that can be advanced for this is that limited natural resources promote early industrialisation which triggers earlier urbanisation. People who migrate from villages into an urban environment become more enterprising, and better functioning markets develop. Savings are then repatriated into the poorer indigenous regions, thereby increasing the social capital of the region.

An abundance of natural resources not only stimulates dysfunctional economic policy choices, but can also pervert political and social behaviour, leading to conflict over the distribution (and non-distribution) of wealth. Countries usually seek to avoid this by using state machinery to bring resources within seemingly rational political control, with Auty (2001b) suggesting strong and transparent governmental involvement is needed in the production of oil. Royalty or taxation policy should guarantee income to the state from oil production. The establishment of an oil fund investment should ideally be a primary requirement to convert a curse into a blessing. Transparent and judicious involvement of the state in the oil sector strengthens one's expectations of benefits one may receive. However, at times, dissatisfaction contributes to reduced political trust in leaders and results in weaker institutional capacities.

An engaged and enlightened private sector, helping promote economic and social diversification, is critical. Hence, building trust through a deep and on-going relationship with the private sector, communities and local governments becomes paramount. Governments and corporations need to co-operate in reducing economic and social costs in order to deliver economic and social benefits. To this end a strong private sector must communicate its integrity in a transparent way. Corporations may seek innovative ways to increase the social and economic benefits that accrue to communities, thereby helping to raise the standards and capacity of public involvement in governance (an absence of which is often a causal mechanism behind conflicts).

Generally it is assumed that an abundance of oil revenue causes broad-based socio-economic and political problems (Ross, 2006). Other authors including Engeli and Pieth (2000) and Wegenast (2007) blame abundance directly for motivating rebellion and allowing the finance of large-scale armed violence. Using a host of alternative measures of natural capital wealth (aggregated as renewable and non-renewable), Soysa (2002) finds that an abundance of renewable resources - not their scarcity - leads to violence and to lower economic, human and institutional development. According to Wagenast (2007), international sanctions for poor governance on all fronts are thus important in ensuring acceptable governmental performance levels in line with the expectations of the different parties. These expectations have to be managed and in this regard the private sector must contribute seriously. A meaningful partnership needs to be encouraged ensuring a fair deal for all stakeholders. Finally, poor economic performance by resource rich countries may be attributed to a lack of democracy. Many authors, including Auty (2001) and Ross (2005), have identified a negative correlation between oligarchies and democratic rule, arguing a democratic state certainly does not ensure good governance for the poor. The Nobel Laureate Amartya Sen holds another point of view. His work on poverty eradication and welfare economics suggest that the way most governments measure poverty by basing it on income may be a flawed perception of well-being.

In their seminal work: *Harnessing windfall revenues: optimal policies for resource-rich developing economies,* Van der Ploeg and Venables (2011) suggest that a windfall of natural resources presents government with an opportunity set of how to manage resources relative to public debt, investment and the distribution of funds for consumption. Generally, the permanent income hypothesis suggests a sustained increase in consumption will be supported by interest on accumulated foreign assets, once resources are depleted. This strategy is not an optimal solution for capital-scarce developing economies. Following this trend of thought, Van der Ploeg and Venables (2011:1) conclude that incremental consumption should be skewed towards present generations and that savings should be directed to accumulation of domestic private and public capital rather than foreign assets. Optimal policy depends on the impact of distortionary taxation and ability of consumers to borrow against future revenues. Van der Ploeg and Venables (2011:2) provide a rigorous analysis of how should a temporary windfall of foreign exchange (from natural resource revenues or foreign aid) be managed by the recipient government? In addressing this issue: they recommend three broad choices:

Firstly, the amount of the windfall that should be saved and what should be the optimal time profile of consumption from the windfall;

The second concerns the choice of whether to invest in the domestic economy or in foreign assets (by increasing foreign exchange reserves or creating a sovereign wealth fund, SWF); and

Third concerns the balance between private or public domestic investment.

The solutions to these issues are dynamic and country specific. To analyse these choices Van der Ploeg and Venables (2011:3) derive policy rules for a welfare-maximising government that experiences a temporary windfall of foreign exchange. They build a family of models in increasingly complex economic environments and derive optimal time profiles for consumption, foreign debt / assets, public investment and tax and transfer policies.

2.7 Kuwait within this context

Much of the discussion in the preceding sections, while relating to several countries outside the Gulf Cooperation Council (GCC), nevertheless highlights issues that the six GCC countries need to consider. While the six countries of the GCC are similar in terms of culture and language they are very dissimilar in terms of governance. The forms of government range from autocracy to democracy. The dramatic growth in state oil revenues has tended to keep the GCC countries insulated from the ravages of the economic crises that have afflicted other resource rich countries. Therefore this section alludes to the resource curse within the GCC, and within Kuwait in particular to determine whether Kuwait is in any way afflicted by the resource curse phenomenon.

In a recent article related to natural resource dependence in the United Arab Emirates (UAE), Karnick and Fernandes (2009) empirically illustrate that the UAE is indeed dependent on the oil sector (as much as any other Gulf country). Since the countries of the GCC are linked to each other and to the rest of the world by geopolitical, social and economic relationships, the findings of Karnick and Fernandes (2009) may theoretically apply to other countries within the GCC. However, each Gulf country has its own individual levels of independence and autonomy. A combination of these levels defines the identity, independence, and external influences upon the country. The Gulf countries thus exhibit great differences as well as similarities. Their identities differ, and they are rivals for autonomy, independence and influence. This rivalry underpins the tendency to compete for oil revenues as the GCC exporters are fully aware of their importance in the world markets. For the oil exporting GCC countries, dependence on the world oil market and oil revenues is very high. Oil revenues support the societies and their differences. However, they realise they are not exclusive and dominant sources of current and future oil or energy supplies.

From a social relations viewpoint, GCC vary in terms of acceptable international norms of democracy and human rights. The challenge that faces Kuwait in particular is how to balance its survival, the prevalence of peace, and the demands placed on it from within and outside its borders. Kuwait needs to broaden its relations beyond oil and energy. Relationships should include other areas of development, namely, political, social and economic. As Kuwait has limited natural capabilities, its dependence on oil is exceedingly high compared to the rest of the GCC. However, its geographical location presents several opportunities (as it is situated between Asia and Europe, and is at the uppermost end of the Arabian Gulf).

Whilst Kuwait is endowed with a rich natural resource in the form of oil, the resource curse as highlighted represents an enormous impediment to real growth and development. It is nevertheless important to understand that an abundance of a natural resource is not the primary problem; but rather, might be the absence of good governance and democratic institutions and structures. In addressing these concerns one needs to consider the nature of political, social and economic institutions and how they have evolved and continue to do so. Different resource rich countries present different profiles in terms of wealth and other economic metrics. In order to understand the resource curse in the Kuwaiti context requires the study of how economic factors have shaped the governance structures of Kuwait and how, in turn, these governance structures influence growth, development and the prosperity of the nation. Income and wealth generated from natural resources ought to be used to support further growth and development, but this has not always been the case in Kuwait and the other GCC countries. This is in part attributable to income being misappropriated by unscrupulous agents. As a consequence the next chapter specifically considers the governance of Kuwait in relation to economic, political and social issues. However at this stage a few macro-economic observations may be made to clarify the situation.

Preliminary discussions with key stakeholders in Kuwait substantiate the widely held belief that Kuwaitis are indeed blessed to have oil within their country. These stakeholders believe that the benefits accruing include: poverty alleviation; infrastructure improvements; enhanced economic growth; job creation, and increasing government initiatives associated with the transfer of technology, innovation and the incubation of related industries. While it is not an intention to dwell upon the views of the different stakeholders at this point, it would suffice to say that their views shape the thinking of governments and private enterprises.

If a resource boom is to be identified as beneficial or detrimental, what matters most is the economic and social influence, therefore how oil wealth is transforming Kuwait needs to be further examined in the light of the changes that taking place in other resource rich countries.

2.8 Resource curse and governance

Heuty and Carlitz (2008) conducting the Open Budget Survey of 2008 reflecting a comprehensive evaluation of budget transparency in 85 countries—finds that resource-dependent countries tend to be less transparent than countries that are not resource dependent. A closer look at the data reveals that the poor performance of resource-dependent countries is largely driven by the lack of budget transparency and accountability in the 22 countries that are considered significant oil and gas producers. Table 2.1 presents a comparative description of various resource rich countries. While the degenerative state of economies in quadrant three (Table 2.1) can in no way be compared to Kuwait, the country still aspires to reach the governance standards evident in developed economies such as Australia and Norway (Quadrant 4).

Table 2.1 Comparative description of various resource rich countries

<table>
<tr><th></th><th>POOR GOVERNANCE</th><th>GOOD GOVERNANCE</th></tr>
<tr><td>MINERAL RICH</td><td>(1)
Zimbabwe (Poor leadership)
Lebanon (War ravaged)</td><td>(2)
Spain (Clean governance)
South Africa (Transparency)
Malaysia (Social values)
Sweden (Clean governance)
Botswana (Clean governance)</td></tr>
<tr><td rowspan="2">OIL RICH</td><td>(3)
Nigeria (Ethnic conflict)
Libya (Autocratic leadership)
Sudan (Ethnic conflict)
Iran (Leadership)
Iraq (War ravaged)
Cambodia (Civil conflict)</td><td>(4)
Norway
Australia
China</td></tr>
<tr><td colspan="2">Kuwait</td></tr>
</table>

Source: Adapted from Raghuram, R. G. and Subramanian, A. (2005) What Undermines Aid's Impact on Growth? Working Paper, No. 05/126, International Monetary Fund, Washington, DC.

The Kuwait government is committed to ensuring good and clean governance. This is considered in the subsequent chapter. However, many issues need to be addressed to ensure that Kuwait progresses permanently into quadrant four.

Each of the countries would be discussed in terms of the four identifiable quadrants. Much of the material employed is derived from two sources: firstly, the Organization for Economic Cooperation and Development - an organization that acts as a meeting ground for 30 developed countries which believe strongly in the free market system and secondly, Secondly, Global Witness (www.globalwitness.org), who run pioneering campaigns against natural resource-related conflict and government corruption (and associated environmental and human rights abuses). Global Witness has exposed the injustices that result from the fight to access and control natural resource wealth, and have sought to bring the perpetrators of this corruption and conflict to book. Their work, like academic authors ranging from De Soysa (2002) and Heuty and Carlitz (2008) to Frankel (2010) has revealed how, rather than benefiting a country's citizens, abundant timber, diamonds, minerals, oil and other natural resources can incentivize corruption, reduce accountability, undermine governments, and even lead to war.

2.8.1 Countries manifesting rich mineral wealth and poor governance

Two countries, namely Lebanon and Zimbabwe are mineral rich economies that suffer from poor governance. Lebanon has a long history of political volatility – which had an adverse impact on investment. Caught in the middle of several different conflicts, Lebanon has paid a high price for the region's instability in the form of lost investment opportunities. There are deposits of high-grade iron ore and lignite; building-stone quarries; high-quality sand, suitable for glass manufacture; and lime. However, Dr. NizarAtrissi, a former Vice President of the Investment Development Authority of Lebanon speaking at an OECD forum and addressing the issue of "Mobilizing Investment for Development in the MENA Region", suggested that while Lebanon has made important strides in structural reform, the promotion of FDI was still lagging behind. He further noted that the current amounts of foreign investment did not meet the growth requirements of the Lebanese economy. Moreover, Lebanon was over-reliant upon short term investments due to the political scenario, failing to attract longer term more stable and beneficial international capital flows. The OECD has identified the challenges facing Lebanon as related to the:

Lengthy bureaucratic procedures impede foreign investment, growth and development in Lebanon. Lebanon is also unable to maintain monetary stability (thus unable to reassure investors of the strength of the Lebanese economy).

Zimbabwe is also identified as a mineral rich-poor governed country. The country has granite (used for building and construction), coal, copper, asbestos and platinum deposits. However, Zimbabwe has been plunged into near anarchy by its President who has held office for more than three decades. An overt and unchallenged manifestation of greed, power and irresponsible government is widely acknowledged to bedevil Zimbabwe, despite it having been a wealthy colonial state. As many Zimbabweans currently lack a stake in Zimbabwe's political institutions, these political institutions have continued to impede the development of genuine governance (Cowen, 2007).

Rich mineral deposits within weak and poor states typically then has a negative effect on their political and economic development.

2.8.2 Countries manifesting Oil-rich wealth and poor governance

Nigeria, Sudan and Libya are oil-rich countries that are situated in Africa whose oil wealth has raised the stakes of political conflict. In such countries an intensive (and sometimes violent) struggle for resource opportunities, has involved inter- and intra- communal/ethnic conflicts, theft and illicit trading in refined and crude oil. De Soysa (2002) examined the interfaces between the Nigerian state, multi-national oil companies, the international community and youth militias within the economy. Although he found that the resource has not caused open violent conflict, it has provided a means for sustaining challenges to state power. The oil economy has helped engender an extensive proliferation of arms the pervasiveness of corruption at the national level – and crime, violence and communal/ethnic conflicts at the local level. Sudan also falls into the same category as Nigeria. Some scholars including Ross (2001) suggest that the Libya's oil wealth helps explain its failure to democratize. He examines how "oil impedes democracy". Libya's resource wealth enabled the government to strengthen its internal security forces and hence repress popular movements and frustrate "modernization effects" (in the sense that growth - based on the export of oil and minerals - failed to bring about the social and cultural changes that tend to produce democratic government). The question that can be asked is whether oil wealth has a consistently antidemocratic effect on states, once other factors are accounted for. Moreover, can this claim be generalized? Whilst oil wealth does not necessarily create autocratic regimes, oil wealth has played a crucial role in sustaining them and in warding off pressures for political change in countries like Nigeria, Libya and Sudan.

Kuwait, a Gulf State, is Iraq's immediate neighbor. Both Iran and Iraq are two Middle Eastern countries that have for long been in the news. Iraq is still recovering from the US liberation that overthrew President Saddam Hussein's regime. However, prior to this the Iran-Iraq war, the Iraqi invasion of Kuwait and punishing economic sanctions and embargos had contributed to a marked deterioration in the Iraqi economy, its infrastructure, and society. Presently, the country is considered to be ungovernable (Paya, 2010). Iraq's neighbor, Iran, has had limited dealings with Iraq since the Iran-Iraq war - described as the Third World's first Great War (Moshaver, 2003). According to Workman (1991) there remains a rooted cultural enmity between Iran and Iraq. This isolation from its nearest neighbor has incided negatively upon a country that underwent an Islamic revolution in 1979, followed by the large-scale "Islamization" of society. Over the past three decades, there has been a lively debate among social scientists about the nature – and consequences - of such developments, and particularly over their implications for other Middle Eastern and Islamic societies (Paya, 2010). The International Monetary Fund (2000) reported that Iran's business and investment environment is governed by unclear and unpredictable laws. Subsequently in 2005 the International Monetary Fund Article IV Consultation report on Iran, suggests that investor protection is weak, as measured by an index including different aspects of corporate governance.

Finally, Cambodia has emerged from genocide and decades of civil war. Cambodia's discovery of oil raised hopes of faster development for the country – but also fears that the "resource curse" might strike again. Global Witness have expressed concern in this regard, reporting that oil exploration licences had been granted largely in secret, while the mining sites it investigated were all owned by the country's political and military elite. Furthermore, the public has not been consulted about the drafting of new laws governing extraction of natural resources. Perhaps the lack of democratic values and accountability is not surprising in a country where the Prime Minister has been in power for almost three decades – despite grassroots organisations pushing for Cambodia's oil to be managed in ways that allow citizens to see what the government does with the resulting revenues – and to hold it and oil companies accountable for their actions.

2.8.3 Countries manifesting rich mineral wealth and good governance

Five countries have been identified for discussion in this category, two European countries, one Asian country and two African countries. Since ancient days, mining and the iron industry have been of great importance in the economic life of Sweden, which was among the most active mining countries in Europe. Iron and steel comprised Sweden's most important industry, and one-third of Sweden's exports are supplied by steel (the country accounts for a large percentage of Western Europe's iron output, and is also home to the region's largest gold mine) (Weil, Gotshal, and Manges, 2002). Regulation and accountability regarding the sector is provided by The Mining Inspectorate of Sweden (Bergsstaten), the agency responsible for the administration of mineral resources in Sweden, housed in the Ministry of Industry, Employment and Communications. The director of the Inspectorate is the Chief Mining Inspector, who is appointed by the Government. The Swedish system of government entails that all power proceeds from the people, and the Swedish governance framework is characterized by several features distinguishing it from that of many other countries - mainly in terms of its long tradition of self-regulation, its structure of corporate governance, and its concentrated ownership dispersion. Regarding the legal and regulatory framework, Weil, Gotshal, and Manges (2002) note that Swedish legislations (particularly the Companies Act) and listing rules demand detailed corporate governance requirements. More recently, the Swedish Code of Corporate Governance was enacted in July 2005 on a “comply or explain” basis as a supplement to the Companies Act and other legislation. Its purpose was to improve overall corporate governance through self-regulation and to address issues of poor corporate governance and decreasing investor confidence. In introducing these strict and accountable standards he Swedish government thus ensures that the private sector exploits the country’s natural resources in a sustainable and transparent manner.

The economy of Spain is the twelfth-largest economy in the world, the fifth-largest in Europe, and the country is regarded as the world's 15th most developed country (Newman 2003). Spain is richly endowed with a vast number of minerals ranging from aluminum to copper, gold, and zinc. According to Newman (2003) Spain has some of the most mineralized territory in Western Europe and - in terms of value of mine output of metallic and nonmetallic minerals and quarry products - Spain was one of the leading European Union (EU) producer countries. Legislation to abolish state and private monopolies was passed in 2002, and the Government continued with its program of liberalizing Spanish industries in 2003. Electricity markets, the natural gas sector, and the petroleum sector continued to be key targets in the liberalization efforts. The natural gas liberalization process has been faster than that required by the EU. (U.S. Energy Information Administration, 2004).Moreover, the U.S. Energy Information Administration (2004) adds that Spain is expected to continue with its privatization and liberalization efforts and Spain’s international profile as a well governed economy has grown appreciably in recent years. This is also consistent with the findings of Newman (2003).

Malaysia is endowed with fertile land and rain forest, with large deposits of tin found in several Malay states. The country is the world's largest producer of tin, rubber, and palm oil. These three commodities along with other raw materials firmly dictated Malaysian economic growth rates until well into the 20th century. As Malaysia moved towards independence, the government began implementing economic five-year plans, beginning with the First Malayan Five Year Plan in 1955. These five-year plans have built in governance standards promoting growth and development. Vandergeest and Peluso (2006) have addressed the issue of sustainability of Malaysia's natural resources in their seminal work: "Forest governance in Malaysia - an NGO perspective". They identify a holistic approach to governance of natural resources and suggest participation (by both men and women) is a key cornerstone of good governance. Adherence to a strict rule of law, with enforcement of that law idone in a transparent manner that abides by the various rules and regulations, ensures that all Malays feel that they have a stake in the ownership and exploitation of the country's forest resources. Accountability is considered a key requirement of good governance in Malaysia, with the Malaysian governmental institutions, the private sector and civil society organizations accountable to both the public and to their institutional stakeholders. Accountability, however, cannot be enforced without transparency and the rule of law.

South Africa with rich agricultural farmlands, rich natural resources and highly developed irrigation systems is a net exporter of food crops. It is also the world's largest producer and exporter of platinum, a significant producer of gold, manganese, chrome, vanadium, and titanium, and also exports a significant amount of coal. In 2000, platinum overtook gold as the country's largest foreign exchange earner. To counter excessive resource dependency the country has supported the (value-added) processing of these minerals to produce ferroalloys, stainless steels, and similar products and sought to diversify – becoming a world leader in several specialized sectors (including motor vehicles and parts, railway rolling stock, and synthetic fuels) (Van Schalkwyk, 2008). Internationally South Africa is also known for its corporate social responsibility (CSR) programmes (Posnikoff, 1997). Recently, however, there have been important developments involving the broadening of the interpretation of CSR and an increasing commitment to the alleviation of poverty and other social ills. Though market-based incentives have contributed to this, the key driver has been the State's legislated transformation programme premised on State sovereignty over mineral resources. (Van Schalkwyk, 2008) Hence, while the interrelationship between companies, state and the institutional context has, in the past, fostered a degree of irresponsibility and minimal collaboration, this has been reversed. In addition, South Africa's post-apartheid governments have made remarkable progress in consolidating the nation's peaceful transition to democracy.

Finally, Botswana which is located in Africa has, according to the IMF, avoided the resource curse. This is curious as, according to Boschini, Pettersson and Roine (2005), the presence of diamonds has arguably been disastrous for the development of countries like Sierra Leone, Liberia and the Democratic Republic of Congo. However, this does not seem be the case in Botswana, which has defied the odds and has a thriving economy. The presence of natural resources, specifically diamonds, played a key role in this accomplishment, according to Lewin (2011). He contends that in Botswana's case, the key to successfully harnessing natural resource revenues lay in good governance and good policies, particularly prudent fiscal and monetary policies, which led to the accumulation of foreign reserves and prevented the volatility that typifies many resource-based economies. Resource revenues supported investments in human and physical capital and improvements in infrastructure which have also raised Botswana's productivity. Like Malaysia, Botswana's non-government organizations play a critical role in the governance of its natural resources.

Lewin (2011) presents some fascinating success stories related to Botswana. In the first few years of independence from the British (1966), there were only 12 kilometers of paved roads and by 2007 Botswana had 7,000 kilometers of paved road and its per capita income had risen sharply making Botswana an upper-middle-income country comparable to Chile or Argentina. Its success is also evident in other measures of human development. At independence, life expectancy at birth was 37 years. By 1990 it was 60, 10 years above the African average. Under-five mortality had fallen to about 45 per 1,000 in 1990, compared with 180 for Africa as a whole. According to the UNDP report by 2011 Botswana's infant mortality rate was 12.59 per 1,000 and life expectancy is 60.93 years. Development assistance has shrunk to less than 3 percent of the government budget. Major strides have also been made in infrastructure and education. Leith (2005) agrees with Lewin (2011), that the discovery and subsequent exploitation of minerals, principally diamonds—was clearly an element of Botswana‘s success.

2.8.4 Countries manifesting oil-rich wealth and good governance

The economy of Norway is best characterized as a developed mixed economy, but with heavy state-ownership in strategic areas of the economy, while the public sector is among the largest in the world (as a percentage of the overall gross domestic product). The country has a very high standard of living compared with many countries, and a strongly integrated welfare system which is among the best in Europe. This success has seen the ideological divide between socialist and non-socialist views on public ownership decrease over time.

Norway was prompt to assert sovereign rights over natural resources in the North Sea, and much of Norway's more recent economic growth has been fueled by an abundance of natural resources, including petroleum exploration and production, hydroelectric power, and fisheries. Agriculture and traditional heavy manufacturing have suffered relative decline – as the 'Dutch disease' syndrome would suggest - compared to services and oil-related industries, However in the last two decades national and local government incentives have been introduced to encourage formation of new mainland industries that are competitive internationally. The Norwegian government has sought to reduce its ownership over companies that require access to private capital markets, and there is an increasing emphasis on government facilitating entrepreneurship rather than controlling (or restricting) capital formation. A transparent approach to the utilization of Norwegian natural resources is at the forefront of good governance and sustainability. The Norwegian experience suggests that there are important linkages between the exploitation of natural resources, environmental issues, and decisions of local, regional and national relevance and - with proper governance - a stronger economy will emerge.

Australia has abundant and diverse natural resources that attract high levels of foreign investment and include extensive reserves of coal, iron ore, copper, gold, natural gas, uranium, and renewable energy sources (Garnaut, 2010). The country is also a significant exporter of natural resources, energy, and food. Key tenets of Australia's trade policy include support for open trade. This policy and the continued demand for commodities, especially from its main trading partner - China - helped the Australian economy to secure positive growth levels. Sustainable natural resource use and management exert pressure on governance arrangements. Jennings and Moore (2000) in their work:*The rhetoric behind regionalization in Australian natural resource management: Myth, reality and moving forward* suggests that governance principles have been developed for the diverse contexts that exist in Australia and, while developed in an Australian multi-level context, these principles can have a wider applicability and significance. They call for governance institutions that are legitimate, transparent, accountable, inclusive and fair.

China, with over 1.3 billion people, is the world's second largest economy after the United States. Like India it is one of the world's fast-growing major economies, with average growth rates of 10% for the past 30 years Nonini (2008). China is the largest exporter and second largest importer of goods in the world, and became the world's top manufacturer in 2011, surpassing the United States. Nonini (2008) also points out that China is a richly endowed country with an abundance of natural resources and oil in particular. China is also the largest producer and consumer of coal in the world. Both critics and admirers of China's economic reform would agree that China is witnessing one of the most extraordinary episodes of social and economic transformation. At the same time Coxhead (2006) contends that rapid resource depletion and reduced industrial growth could expose China to a period of slow economic growth, as predicted by the ''natural resource curse'' hypothesis. Coxhead (2006:1) goes on add "that the idea of a resource curse has never resonated loudly in South East Asia a resource-abundant region whose largest economies have recorded exceptionally high growth rates of per capita income since the 1970s".

The political incentives that natural resources generate are the key to understanding whether or not they are a curse. Countries with institutions that promote accountability and state competence will tend to benefit from resource booms and countries - without such institutions however may suffer from a resource curse. While evidence was provided in this chapter to show where natural resource abundance is associated with various negative development outcomes, the chapter also shows that abundance does not necessarily 'curse' a country – providing appropriate governance structures (budget transparency and public accountability – though not limited just to these two components) are present. Explanations for the resource curse (or how it has been circumvented) therefore need to consider the role of social forces and the domestic political and economic environments that are shaping governance structures in countries like Kuwait.

2.8.5 Kuwait and governance

The question therefore arises as to where would one locate Kuwait in Table 2.1. However, it would suffice at this juncture to say that Kuwait lies somewhere between quadrants 3 and 4 for while it certainly does not betray the characteristics of poor governance that are seen in oil-rich economies such as Nigeria and Libya, neither has its institutional structures evolved in a manner that reflects those found in countries such as Norway and Australia. (The following chapter examines the institutional structures of Kuwait in more detail to support this contention)

While some governments of resource rich countries are encouraged to step up their efforts to tackle bribery and corruption in order to boost job creation, and improve living standards, the ongoing political upheavals in the Arab world have also highlighted the need for governments to work more closely with civil society, business and labour unions in order to reinforce citizens' participation and trust in government. This is the task that confronts Kuwait, and to this end the following chapter also charts the evolution of political, economic and social forces in Kuwait.

The literature findings of this section suggest that resource curse is a function of many factors or variables and governance is an important variable. Should natural resources in the form of oil as in the case of Kuwait be the hallmark of progress and advancement, poor governance would be its bane resulting in resource curse effects. Consistent with this discussion Van der Ploeg (2011:1) provides cross-country evidence that rejects the traditional interpretation of the natural resource curse. He goes on to suggest:

Growth has a negative relationship with volatility of unanticipated output growth independent of initial income, investment, human capital, trade openness, natural resource dependence and population growth;

The direct positive effect of resources on growth is overcome by the indirect negative effect through volatility;

With well-developed financial sectors, the resource curse is less pronounced. Fourth, landlocked countries with ethnic tensions have higher volatility and lower growth;

Restrictions on the current account raise volatility and depress growth whereas capital account restrictions lower volatility and boost growth.

Van der Ploeg's (2011) goes on to conclude that volatility is a typical feature of the resource curse and Kuwait needs to be considered within this context and this is addressed in the following Chapter.

2.9 Conclusion

This chapter has considered historic and current literature and theories concerning resource curse addressing the following three main issues. Firstly, is an abundance of natural resource necessarily an impediment to development? Secondly, what are the major causes of the resource curse? Thirdly, how can the resource curse be converted to wholesome blessings? Having reviewed the available related literature it has shown that some of the research findings associated with the resource curse relate to anecdotal evidence and therefore the evidence is not unequivocal that resource abundance translates into a cursed economy. In addition, whilst the social consequences encountered are varied, the extensive review identified that some resource-rich countries have performed exceptionally well in terms of economic metrics, and thereby overcome the resource curse. Natural resource wealth fosters different political ideologies which in turn promote development performances that are not consistent across developing economies. In considering the curse – one must ascertain who it really hurts. A key to successful management of income from natural resources is equitable distribution between rich and poor (within states and across regions), with the intention of reducing poverty, ameliorating economic growth and reducing corruption.

CHAPTER THREE

THE POLITICAL AND ECONOMIC DEVELOPMENT FRAMEWORK

"My father rode a camel, I drive a Mercedes, my son flies an airplane.
My grandson, ah! my grandson, he will ride a camel,
more for pleasure than out of necessity."
(Anonymous)

3.1 Introduction

The previous discussions have shown that the governance and association between economic development and oil wealth is contentious. A pressing and perplexing problem throughout the world today is the inability of many countries to convert valuable natural resources into enhanced standards of living for their citizens. Many scholars have begun to look at this problem as a "resource curse," a phenomenon denoting an inverse relationship between endowment with natural resources and economic growth (as examined in Chapter Two). However, "leadership curse" may aptly describe the predicament of many countries. A defining characteristic of many resource-rich countries is the discrepancy between the interest of the stewards of the resources and the true owners of the resources. At times those in political office (the stewards) appear to work extremely hard to ensure that the rest of the population (the owners) receive little benefit from the resources with which their countries have been abundantly endowed, and so the **governance of natural resources** merits further research.

However, an explanation of the development of any country's natural resources is incomplete if one does not consider the impact of the historical, political and social milieu that pervades an economy. Studying the historical development of the economy provides a context for understanding existing resource challenges, and a source of hypotheses to explain the trajectory of its evolution. Economic history has become increasingly important as a framework for studying questions not just in economic history, but also in other fields of economics, particularly macroeconomics, microeconomics and development economics. Empirical research from the Arab Planning Institute in Kuwait supports this claim in the Kuwaiti context (Ali Abdel Gadir, 2009). For any nation that has only come to the world's attention since the discovery of oil, one could assume that in-depth and accurate recording of a flourishing civilization prior to any oil discovery might have been lost. However, much of this rich cultural history can dictate the future path of economic development. Such is the situation of Kuwait. Social and cultural aspects deriving from Kuwait's history, together with political developments, exert a strong influence over the economic activities of contemporary Kuwait. Over the years, intervention in the economy by past rulers resulted in policies ensuring the benevolent distribution of income and wealth. Consequently, this chapter commences with a brief historical survey of Kuwait and its impact on the governance of its natural resources. It addresses the emergent economic issues within these confines, and discusses the role of the government and private sectors in sustaining social stability, economic growth and development of the country. This provides a basis for understanding the dimensions of innovation in Kuwait.

3.2 Social changes in Kuwait from an historical perspective

An understanding about the effects of political institutions on economic growth and development in Kuwait is necessary to show how the patriarchal conduct of past rulers manifested itself in a "sharing and caring" relationship with the citizens, and how this has continued to underpin the relationship between government and Kuwaiti society. A loyalty to leadership in early Islamic society is a manifestation of a leader's loyalty to his citizens according to Mottahedeh (1980) and Afsaruddin (2002) suggests that this loyalty is important in countries like Kuwait.

Although Kuwait has existed for 250 years as an independent state, it can be considered a newcomer in world history. While archaeologists uncovered a wealth of historical treasure and information dating back to 4500 BC., Rush (1987) together with other historians, like Al-Hijji (1997), Crystal (1995), Jamal (1998) and Bondarevsky (2003) date the origins of Kuwait as a nation back to the beginning of the eighteen century when the tribal confederation of Banu Utib migrated to the Gulf from the Najd region in Central Arabia in order to escape drought and famine. After a stay in Basra, the Utibs wandered south to settle on the coast of the Gulf, at the present day location of Kuwait City. Here, they established a self-governing political unit combining a concern with inequality with self-determination for the newly settled citizen. The historians already cited, contend that the modern political history of Kuwait began around this time and, for convention, the year 1752 was denoted as the year these settlers elected an Al Sabah Sheik as their leader. Since then rulers have traditionally come from the Al Sabah family tree. The leader was chosen by a family council, in consultation with the leading families who, together with the tribal elite, have exercised some restriction over the Sheik's political authority.

The task of leading a tribe during this time was basically to represent one's community in negotiations with Ottoman Turkey or neighbouring tribes. According to Rush (1987), a major - albeit unsuccessful - challenge to this newly established system of rule occurred when the Al Khalifa family fell out with the Al Sabah family. As a consequence the Al Khalifa family left Kuwait for Qatar in the 1760s moving onto Bahrain, where they continue to rule. Despite this, the two families maintained close economic links and, as early as 1770, Kuwait handled most of the trade in the Gulf (including transportation to India), and was the pivot of the overland trade route to the Mediterranean.

It is widely reported in literature (Rush, 1987) that, during the nineteenth century, members of the Al Sabah ruling family controlled the growing trade in pearling in Kuwait. The Al Sabah family also developed a trading relationship with Britain, firstly through the British East India Company in 1775. Moreover, being members of a small, vulnerable settlement, the Al Sabah family sought to maintain cordial relationships with not only the British, but also the Wahhabis of Arabia. It was during Sheik Abdellah Al Sabah II's rule (1866 – 1892), that Kuwait developed close ties with the Ottomans. In 1871 he took the Ottoman title of provincial governor (*qaimaqam*). Kuwait's domestic politics remained unaffected by these closer ties, however, because the Ottoman government was simply interested in using the Gulf as a highway from Mesopotamia to India at the time (Abu-Hakima, 1983).

Kuwait moved more fully into the British sphere of influence at the end of the nineteenth century when Sheik Mohammed Al Sabah's son, Mubarak the Great (1896-1915) requested British military support. The Ottomans were supporting allies of Mubarak's brothers (Kuwait's previous rulers) whom Mubarak had killed on taking power in 1896. Britain at this time was concerned with growing European involvement in the Gulf, most notably as the Ottomans had granted a concession to Germany for the construction of a Berlin-to-Baghdad railroad (with a further proposed line to Kuwait). Britain therefore signed a treaty in 1899 promising British military support in return for control of Kuwait's foreign policy. This treaty remained operative until Kuwait's independence in 1961. Commerce continued with Kuwait remaining a key Gulf trading post, importing textiles, rice, coffee, sugar, wheat, tobacco, fruits, spices, teak and mangrove. Ships docked in Kuwait as they passed through the Gulf, sailing east to India and west to Africa, while there were caravan links to the interior of the Arabian Peninsula.

Trade had a number of effects upon Kuwaiti society. First, by bringing locals into contact with people from all over the world, it gave Kuwaitis a more cosmopolitan outlook. Secondly, as Kuwaiti men became involved in trade, home life was strictly dominated by "resourceful and self reliant women" (Al-Hijji, 1997). This trading prowess also introduced Kuwaitis to contracting, finance and investment, experiences and skills that have helped the development of modern Kuwait.

BOX 3.1 THE RULERS OF THE AL-SABAH DYNASTY

- Sheikh Sabah I bin Jaber (1752 - 1762)
- Sheikh Abdullah I Al-Sabah (1762 - 3 May 1814)
- Sheikh Jaber I Al-Sabah (1814 - 1859)
- Sheikh Sabah II Al-Sabah (1859 - Nov 1866)
- Sheikh Abdullah II Al-Sabah (Nov 1866 - May 1892)
- Sheikh Muhammad Al-Sabah (May 1892 - 17 May 1896), assassinated by successor
- Sheikh Mubarak Al-Sabah, "the Great" (18 May 1896 - 28 Nov 1915)
- Sheikh Jaber II Al-Sabah (28 Nov 1915 - 5 Feb 1917)
- Sheikh Salim Al-Mubarak Al-Sabah (5 Feb 1917 - 22 Feb 1921)
- Sheikh Ahmad Al-Jaber Al-Sabah (22 Feb 1921 - 29 Jan 1950)
- Sheikh Abdullah III Al-Salim Al-Sabah (29 Jan 1950 - 24 Nov 1965 became Amir on 19 Jun 1961)
- Amir Sabah III Al-Salim Al-Sabah (24 Nov 1965 - 31 Dec 1977)
- Amir Jaber III Al-Ahmad Al-Jaber Al-Sabah (31 Dec 1977 - 15 Jan 2006)
- Amir Saad I Al-Abdullah Al-Salim Al-Sabah (15 Jan 2006 - 24 Jan 2006), (due to illness)
- Amir Sabah IV Al-Ahmad Al-Jaber Al-Sabah (29 Jan 2006 - now)

Source: Rush (1987)

After Sheik Mubarak's death, Kuwait was ruled by Sheik Jaber Al Sabah (1915-17) and Sheik Salim Al Sabah (1917-21), two of his sons. Thereafter, only descendants of these two sons would rule Kuwait. After

Sheik Salim's death (1921), Kuwait was governed for almost thirty years by Sheik Ahmad Al Jaber Al Sabah. However, the Great Depression, a decline of the pearling industry, and a trade dispute with Saudi Arabia that culminated in a Saudi embargo against Kuwait, triggered a serious recession. In 1938, an unsuccessful rebellion took place, the same year that the Anglo-American Kuwait Oil Company found oil in the Burgan oil field. Although World War II prevented immediate exploitation of the find, Kuwait made its first international shipment of oil in 1946.

Simultaneously, oil concessions were signed with Britain which promised Kuwaitis better times. To prevent the resulting income being sequestrated by the ruling family, however, the leading merchants petitioned Sheikh Ahmad Al Jaber Al Sabah for greater political and economic freedom.

Sheik Ahmad Al Jaber was succeeded by his cousin Sheik Abdellah, who is Sheik Salim Al Sabah's son (1950 - 65), took personal charge of the distribution of the oil revenues, the consequence of which was the emergence of a large bureaucratic state as Kuwait became a wealthy oil Sheikdom. Sheik Abdellah also oversaw Kuwait's transformation into a formally independent state (June 19, 1961) as he signed new letters of friendship with the British to replace the treaty of 1899.

Domestically, Sheik Abdellah Salim made two key political decisions. The first was to distribute oil revenues more widely among the population, notably through the provision of education and health care. The second was to permit the formation of a newly elected National Assembly, the first elections for which were held in 1963.

Upon Sheik Abdellah Salim's death (1965), his brother Sabah - as Salim Al Sabah - took power and continued and consolidated the policies set in motion by his predecessor. When Sheik SabahSalim died in December 1977, Sheik Jaber Al Ahmad Al Jaber Al Sabah (1977-2006) became the new Amir.

Several external events dominated Sheik Jaber Al Ahmad's rule. Firstly, the Iran-Iraq War, which triggered an increase in the level of political violence in Kuwait. Secondly, the 1983 bombing of the United States Embassy and the 1985 public assassination attempt on the Amir. Tensions associated with the war also exacerbated divisions in Kuwaiti society, and led the Amir to restrict public participation in political life. Thirdly, the Iraqi invasion of Kuwait (August 1990) which placed Kuwait under direct foreign rule with sovereignty being restored in February 1991.

Since 1991 and the restoration of sovereignty, Kuwait's oil and (oil related industries) have driven Kuwait's economy (as is clear from the following section of this chapter). Understandably, oil was seen as a national asset and contemporary rulers followed the lead of their predecessors in seeing the need for this natural resource wealth to be shared by the nation. Thus as Kuwait prospered nationally, so has its people benefited individually. If one accepts that the sea had shaped the past economic history of Kuwait and oil is shaping its present, the governance of this natural resource will be shaping the future economy.

Governance is associated with power, relationships and accountability. It thus has a major influence on the effectiveness of management, and the maintenance of community and environment (sustainability). Hence, the rulers of the Al-Sabah dynasty, in conjunction with the national constitution, have an important role to play in reconciling these objectives.

3.3 Constitutional imperatives driving change

New constitutions are needed to cope with modern economic challenges. Modern institutions and constitutional political economy have contributed significantly to an understanding of how legal and political institutions interact with economic processes; more specifically, how different constitutional arrangements affect and are affected by the development and regulation of markets. No serious analysis of the process of economic development is possible without acknowledging the nature and functioning of constitutions. Besides theoretical insights, numerous cross-country studies (refer Raghuram and Subramanian, 2005) have provided empirical evidence on the (economic) effects of various characteristics of national constitutions such as human rights, judicial independence, competition policy, direct democracy, supreme audit and central bank independence. Sachs and Warner's (1995a; 1995b; and 1997) are widely-cited studies in this regard. Sarraf and Jiwanji (2001) also reflect upon the situation in Botswana whilst Ali (2009) considered the Arab region. Constitutional political economy not only assesses the economic consequences of these constitutional arrangements, but strives to explain its interactions with other constitutional players, the legislative and the executive branch. This section focuses on the role of the constitution (its effectiveness) in economic and social regulation, taxation and redistribution. The institutional framework of the Kuwaiti economy bears the mark of the constitution.

Following independence in 1961, and under an Iraqi threat of occupation, Amir Abdellah Al Salim announced that he would establish a constitution for Kuwait. Elections were held for a Constituent Assembly, which then drafted a constitution (promulgated as Law Number 1 on November 11, 1962). Although articles of the constitution have since been suspended twice (in 1976 and 1986), the document nonetheless remains the basic statement of intent underpinning the Kuwaiti political and economic systems.

The constitution defines Kuwait as "an independent sovereign Arab State," with its people "part of the Arab Nation." Islam is "the religion of the state," and the Sharia (Islamic law) is "a main source of legislation." The constitution further establishes Kuwait as "a hereditary Amirate, the succession to which shall be in the descendants of the late Mubarak Al Sabah" (Baklini, 1982). Furthermore, the constitution grants citizens a number of social rights, which form the basis for Kuwait's extensive welfare system. The state is constitutionally obliged to care for the young and to aid the old, the ill and the disabled. It is obliged to provide public education and to attend to public health. The constitution further provides for state involvement in the national economy to the degree that these obligations are fulfilled (The Library of Congress Country Studies; CIA World Fact book).

Cooperation between public and private actors is an essential characteristic of a constitution that pledges support for economic growth, development, productivity, improvement of the standards of living and prosperity for all citizens, and unites all Kuwaitis towards a common goal. Moreover, according to Rush (1987) unity and justice, long the objectives of former leaders, have become a sacred duty of all Kuwaitis, so as to build a strong and developed Kuwait on the basis of freedom, unity, peace, stability and security. These have become the core social values of not only Kuwait, but other GCC countries too. Vine and Casey (1992) are convinced that Kuwait's constitution recognizes, protects and secures the rights and freedoms of citizens based on human dignity and equality, and that this guarantees a firm constitutional basis for Kuwait's economic growth and development.

3.4 Kuwaiti economic growth and development

In order to place the debate about innovation in context, it is also necessary to understand and interpret the nature of economic growth and development in Kuwait. For the Kuwaiti rulers, growth is based upon two basic inseparable principles – economic growth and social development. Economic indicators make it clear that the Kuwaiti economy witnessed a significant resurgence in 2007-2008 because of the important and well-documented developments in the world's energy markets. These markets recorded an increase in oil prices which translated into very high rates of economic growth. Gross domestic product doubled in a short period of time. As a result of the country's economic reliance on oil as its major driving force, the public sector in Kuwait dominates the economic sphere. Attempts to develop non-oil industries have been negated by the limited domestic market, the absence of other natural resources and the resource curse (as outlined in Chapter Two), although the government is currently seeking to develop the manufacturing and financial services sector. Petrochemical industries remain the leading industrial sector and support the production of fertilisers and other chemicals (salt, chlorine, caustic soda and chlorine gas). Other light industry includes food processing, textiles, furniture, paper, mineral and metallic products, cement, sulphur processing, detergents and construction materials.

The government also has plans for major infrastructure projects including development of the pier (at Mina Ahmadi); a new city in the north (Silk City) including a causeway from central Kuwait; a new power station (Al Zoor); a new waste water treatment plant, (Sulabiya); redevelopment of the centre of Kuwait City and other housing projects. As the management has yet to be decided for most of these major projects – the timescale for actual construction is therefore uncertain at present, although they will all be funded from oil reserves.

Healthcare and the environment are also issues that Kuwait is currently concerned about. The health care system has from its inception improved markedly decade on decade. Historical data compiled by the Chief Federal Research Division - Library of Congress indicates that Kuwait first introduced a modern health care system in the first years of the twentieth century when Sheikh Mubarak Al Sabah the Great requested doctors from the Arabian Mission of the Dutch Reformed Church in the United States to establish a clinic. By 1911 the group had established a hospital for men and eight years later a small hospital for women. According to Mortimer (1993) a thirty-four bed Olcott Memorial Hospital opened in 1934. It is not surprising then that the increase in oil revenues resulted in the government expansion of the health care system. The Amiri Hospital opened in 1949, The Kuwait Oil Company (KOC) also opened some small health facilities, and in the 1950s the government introduced a comprehensive health care system (including veterinary services) for the entire population. Expenditures on health ranked third in the national budget, public works being first and education second. By 1990 Kuwait had an extensive health and welfare programme, including: free medical services from highly trained practitioners in modern facilities; free education from primary through to university level; subsidized food, housing, utilities, and transportation (Mortimer, 1993). The system is supported entirely by oil revenues.

To further understand economic growth and development fully, and appreciate how oil has come to dominate Kuwaiti life, one needs to examine the Kuwait economy over six distinct periods, namely: the pre-oil period; 1938 -1962, the discovery of oil; 1963 – 1972, Kuwait becomes a major oil producer; 1973 – 1982, the oil bonanza; 1983 – 1991, oil prices fall and 1992 – today.

3.4.1 Pearling dominates in the pre-oil period

In the pre-oil period, pearls were Kuwait's main natural resource. As a result, shipbuilding, using imported materials, also became an important local industry. Large trading dhows frequently set out for India to return with merchandise which was then loaded onto desert caravans bound for various parts of the Mediterranean. Kuwait served as the gateway to the region because of its strategic geographical location. Caravans from southern and eastern Arabia also passed through Kuwait on their way to Syria. By 1929 Kuwait's pearling industry, which once boasted 800 pearling ships, had almost disappeared following the introduction of Japanese cultured pearls and the worldwide fall in demand for luxury goods occasioned by the Great Depression.

Education during this period was limited to the teaching and reciting of the Holy Quran or the three basic Rs: reading, writing and arithmetic and was consistent with the economic and social conditions at the time. The expansion of trade led to the establishment of the first schools in Kuwait, the Al-Mubarkiya School in 1891 and the Al-Ahmadiya School in 1921. Both schools, however, concentrated merely on arithmetic and religious instruction. A more structured educational system emerged in 1936, when a Council of Education was set up to formulate a more meaningful education policy. During the period, education came to be seen as a tool that would endow men and women with the necessary skills to reduce dependence upon Britain.

3.4.2 Oil begins to flow (1938-62)

The first exploration well was drilled at Bahrah (Al-Hijji, Yacoub, 1997). In February 1938, oil was discovered at Burgan and eight more wells were drilled in the vicinity during the next four years, although it was not until the end of the World War II that the first crude oil shipment was exported. Sheikh Jaber became Chief of Public Security in the Al-Ahmadi oilfields in 1949 and after completing his education in local private schools and from palace tutors, he dealt with the oil companies in various capacities.

In 1959 he became head of the Department of Finance (which became the Ministry of Finance after Kuwaiti independence in 1962). Much of both the development of the oilfield and the democratization of the economy can be attributed to his untiring efforts to make Kuwait a highly ranked producer of oil. It was during his tenure that the country adopted a constitution in 1962 and established a National Assembly holding legislative authority (the Amir and his Council of Ministers holding executive powers). When the President of Iraq, Abdulkarim Qasim, claimed that Kuwait was really Iraqi territory, Abdullah Al-Salem requested Britain send a force to defend Kuwait. The League of Arab States then sent a joint Arab security force to replace the British force, rejected Qasim's claims, and recognized Kuwait's independence. Kuwait joined the Arab League on 20 July 1961, and then the United Nations on 14th May 1963 (Al Mughni, 2001).

The Credit Bank was established by Law 40/1960 to facilitate loans for Kuwaiti citizens using oil revenues. State employees could use their salaries and bonuses as collateral. The Credit Bank was later replaced with the Savings and Loan Bank by Law 30/1965 with a mission to encourage people to save and invest.

The foundation of applied education and training was laid using oil revenues, and with a view to supporting the development of oil exploration, production and export in Kuwait. In the 1950s, the State began to establish training centres and organize programmes to prepare the manpower needed for the oil industry. The Ministry of Education also established a number of specialised tertiary institutions while the other ministries created their own in-house training centres and institutes as well, with the Ministry of Health recruiting Egyptian doctors to train health care professionals. In light of such piecemeal development, the State found it imperative to establish a central body (referred to as the Allied Health Workers' Forum) to supervise and coordinate the activities of these numerous institutes. In 1956, oil revenues enabled the government to introduce a major education plan that divided formal education into four categories: Kindergarten (with a duration of two academic years); Primary (with a duration of four academic years); Intermediate (with a duration of four academic years); and Secondary (with a duration of four academic years). Post-secondary education was also restructured using oil income during this period.

3.4.3 Kuwait: A major oil producer (1963-72)

By the decade of the sixties, Kuwait ranked as a major oil producer with an output of 2.5 million barrels per day, accounting for about 23 percent of area production and 7 percent of world crude oil production by 1967 as reflected in Table 3.1. In that same year, the petroleum industry provided more than 95 percent of all government revenues and foreign exchange earnings, revenues from oil companies operating in Kuwait and in the Kuwait-Saudi Arabia Neutral Zone amounting to about $650 million each year.[2]

The Kuwait Chemical Fertilizer Company began production of ammonium sulphate and continued construction of related facilities to produce urea. These plants had a daily production capacity of 500 tons of ammonium sulphate and 550 tons of urea. Construction of the Kuwait National Petroleum Company (KNPC) oil refinery near Ahmadi proceeded as scheduled and additional petrochemical and other industrial development projects were completed. Under the terms of a joint venture agreement, a Turkish company built a chemical fertilizer plant at Mersin (Turkey) with Kuwait capital, with liquid ammonia being supplied by Kuwait under a long-term contract. Sheikh Jaber was also appointed as Deputy Prime Minister, while remaining Minister of Finance and Industry, thus strengthening his reputation as a vigorous, dynamic figure that shaped the economy.

Table 3.1 shows world crude production between 1962 and 1972 for selected countries, including Kuwait.

[2]The Saudi-Kuwaiti Neutral Zone, also known as the Divided Zone, is an area of 5,770 km^2 between the borders of Saudi Arabia and Kuwait. In 1966 Kuwait and Saudi Arabia agreed to divide the Neutral Zone, the partitioning agreement making each country responsible for administration in its portion was signed in December 1969.

Table 3.1 World crude oil production, (1962-1972)
(Million Barrels per Day)

Year	Persian Gulf Nations	Selected OPEC Producers								World							
		Iran	Iraq	Kuwait [3]	Nigeria	Saudi Arabia	United Arab Emirates	Venezuela	Total OPEC [4]	Canada	China	Mexico	Norway	Former U.S.S.R.	United States	Total Non-OPEC	World
1962	6.19	1.33	1.01	1.96	.07	1.64	.01	3.20	R 10.50	.67	.12	.31	.00	3.67	7.33	13.85	24.35
1963	6.82	1.49	1.16	2.10	.08	1.79	.05	3.25	11.51	.71	.13	.31	.00	4.07	7.54	14.62	26.13
1964	7.61	1.71	1.26	2.30	.12	1.90	.19	3.39	12.98	.75	.18	.32	.00	4.60	7.61	15.20	28.18
1965	8.37	1.91	1.32	2.36	.27	2.21	.28	3.47	R 14.34	.81	.23	.32	.00	4.79	7.80	R 15.99	30.33
1966	9.32	2.13	1.39	2.48	.42	2.60	.36	3.37	R 15.75	.88	.29	.33	.00	5.23	8.30	R 17.20	32.96
1967	9.91	2.60	1.23	2.50	.32	2.81	.38	3.54	R 16.79	.96	.28	.36	.00	5.68	8.81	R 18.60	35.39
1968	10.91	2.84	1.50	2.61	.14	3.04	.50	3.60	R 18.71	1.19	.30	.39	.00	6.08	9.10	R 19.93	38.63
1969	11.95	3.38	1.52	2.77	.54	3.22	.63	3.59	R 20.85	1.13	.48	.46	.00	6.48	9.24	R 20.84	41.70
1970	13.39	3.83	1.55	2.99	1.08	3.80	.78	3.71	R 23.42	1.26	.60	.49	.00	6.99	9.64	R 22.47	45.89
1971	15.77	4.54	1.69	3.20	1.53	4.77	1.06	3.55	R 25.31	1.35	.78	.49	.01	7.48	9.46	R 23.21	48.52
1972	17.54	5.02	1.47	3.28	1.82	6.02	1.20	3.22	R 27.11	1.53	.90	.51	.03	7.89	9.44	R 24.03	51.14

Source: Arab Petroleum Research Center, Arab Oil and Gas Bulletin (2009)

The average quantity of oil produced during this period was 2.587 million of barrels per day, Kuwait recording its highest output of 3.28 million barrels per day in 1972. While Kuwait only produced 5-6% of total global oil output in this period, it was nevertheless the fifth largest producer, after the US, the USSR, Saudi Arabia and Iran.

Following independence in 1961, the leaders of Kuwait developed a strategy for a comprehensive welfare system. The education policy of the country was outlined in an overall plan detailed in the April 1966 Amiri Decree. Following this, the first higher education institution in Kuwait - Kuwait University (KU) – was established with two faculties: the Faculty of Science, Arts and Education and the Girl's College. Studies started in 1966-67 with 418 students and a faculty of 31. In the first batch, 359 were Kuwaitis, 70 per cent of the total number of Kuwaiti students who had passed the secondary school certificate examinations. Females represented 42 per cent of the total number of students enrolled that year.

The Central Bank of Kuwait was established as late as 1969, with a mission to lay the foundations for, and maintain a flexible and stable monetary financial system in the State of Kuwait. Their mandate included; issuing the Kuwaiti Dinar on behalf of the State of Kuwait; directing credit policy as necessary to assist social and economic progress; regulating the banking system in the country, and acting as a banker and financial advisor to the government. Representatives from both the public and private sectors make up the Board of Directors. The Central Bank of Kuwait guides all monetary policies in the country and strictly monitors the exchange rate. The Central Bank also moderates the conduct of Kuwait banking institutions. The Kuwait Fund for Arab Economic Development (KFAED), commonly known as the Kuwait Fund, is the State of Kuwait's aid agency charged with providing financial and technical assistance to developing countries. Founded in December, 1961 by the then Minister of Finance, Sheikh Jaber Al-Ahmad Al-Jaber Al-Sabah, the Kuwait Fund was tasked with ensuring that Kuwait's newly discovered oil wealth benefit neighbours and "friends" alike.

According to Robert McNamara, President of the World Bank (1968-1981) the Kuwait Fund was the first aid agency in the world to be established by a developing country. Originally established with a capital of KWD 50 million, this increased to KWD 200 million in 1966."When first established in 1961, the Kuwait Fund was without precedent. Here was Kuwait, a tiny country, until recently among the poorest places in earth, establishing a development fund in the year of its political independence. While welcoming its new-found prosperity it was declaring a willingness to share its future wealth with its Arab neighbors." (World Bank Archives, 1981).

It is managed as a semi-independent arm of the Kuwaiti government and maintains sovereignty over its operations. In July 1974, the scope of the Fund's activity was extended to the rest of the developing world (from its original mandate of solely helping countries in the Arab World) and its capital increased from KWD 200 million to KWD 1000 million (Al-Hijji,1997).

3.4.4 The oil bonanza (1973-82)

Al-Hijji (1997) indicates that in 1974 the Kuwait National Assembly ratified an agreement giving 60 per cent control of the operations of the KOC to the State of Kuwait (the residual 40 per cent being divided equally between BP and the Gulf Oil Corporation). The following March, the Kuwaiti government took control of the remaining 40 per cent of shares, thus assuming full control of KOC. The period also saw Kuwait move heavily into downstream activities including local refining, transport, overseas refining, and petroleum distribution.

The resulting massive inflow of funds, despite declining national production levels as detailed in Table 3.2, were spent on further developing the country's infrastructure and raising living standards. Roads, ports, factories, power generating stations and desalination plants were constructed and the population increased sharply as thousands of foreign technical advisors and workers arrived to support Kuwaiti development. Kuwaitis now found themselves in the minority, working as importers, contractors, landlords and government officials.

Table 3.2 World crude oil production, (1973 – 1982)

World Crude Oil Production, 1973 - 1982 (Million Barrels per Day)																			
		Selected OPEC Producers								Selected Non-OPEC Producers									
Year	Persian Gulf Nations	Iran	Iraq	Kuwait	Nigeria	Saudi Arabia	United Arab Emirates	Vene-zuela	Total OPEC	Canada	China	Mexico	Norway	Former U.S.S.R.	Russia	United Kingdom	United States	Total Non-OPEC	World
1973	20.67	5.86	2.02	3.02	2.05	7.60	1.53	3.37	R 29.66	1.80	1.09	.47	.03	8.32	--	(s)	9.21	26.02	55.68
1974	21.28	6.02	1.97	2.55	2.26	8.48	1.68	2.98	R 29.32	1.55	1.32	.57	.04	8.91	--	(s)	8.77	26.39	55.72
1975	18.93	5.35	2.26	2.08	1.78	7.08	1.66	2.35	R 25.79	1.43	1.49	.71	.19	9.52	--	.01	8.37	27.04	52.83
1976	21.51	5.88	2.42	2.15	2.07	8.58	1.94	2.29	R 29.12	1.31	1.67	.83	.28	10.06	--	.25	8.13	28.23	57.34
1977	21.73	5.66	2.35	1.97	2.09	9.25	2.00	2.24	R 29.58	1.32	1.87	.98	.28	10.60	--	.77	8.24	30.12	59.71
1978	20.61	5.24	2.56	2.13	1.90	8.30	1.83	2.17	R 28.16	1.32	2.08	1.21	.36	11.11	--	1.08	8.71	32.00	60.16
1979	21.07	3.17	3.48	2.50	2.30	9.53	1.83	2.36	R 29.35	1.50	2.12	1.46	.40	11.38	--	1.57	8.55	33.32	62.67
1980	17.96	1.66	2.51	1.66	2.06	9.90	1.71	2.17	R 25.38	1.44	2.11	1.94	.49	11.71	--	1.62	8.60	34.17	59.56
1981	15.25	1.38	1.00	1.13	1.43	9.82	1.47	2.10	R 21.22	1.29	2.01	2.31	.47	11.85	--	1.81	8.57	34.83	56.05
1982	12.16	2.21	1.01	.82	1.30	6.48	1.25	1.90	R 17.77	1.27	2.05	2.75	.49	11.91	--	2.07	8.65	35.68	53.45

Source: Arab Petroleum Research Center, Arab Oil and Gas Bulletin (2009)

While elections for the first National Assembly were held in 1963, and representatives of the country's leading merchant families won the majority of the seats, the limited success of the democratization experiment caused the Amir to subsequently dissolve the Assembly. It did not reopen until 1981 and then only for a short while until further suspension (reopening again in 1986).

In the educational arena, the Public Authority for Applied Education and Training was established on 28th December 1982 with the objective of supplying manpower to meet the technical manpower shortages created by the rapid industrial and economic development of the country.

3.4.5 Oil prices slump (1983-1991)

In 1983, KPC completed its acquisition of the Gulf Oil Refining Company (United States), an acquisition that meant Kuwait could now market its own oil related products. However, the production and processing of crude oil continued to be the mainstay of the economy during this period, providing further revenues to develop its growing downstream industries (Table 3.4), despite the generally weak level of world oil demand and the oversupply situation that persisted through most of the 1980s. Kuwaiti oil production rebounded slightly from a comparatively bad year in 1982, but crude oil output remained at less than one-third of its 1972 peak of 3.28 million barrels per day (and only 53% of its previous 10 year period average). The average quantity produced during this period was 1.21 million barrels per day as shown in Table 3.3. This comparatively low level of oil production was considered a more severe problem by the Kuwaitis than the general decline in oil revenues. The drop in oil output saw a sharp decline in the associated production of gas, which was used throughout Kuwait for LPG and ammonia-urea fertilizer production, as well as for its water desalinization plants and power generating stations. These facilities were forced to operate well below capacity, reducing export revenues further.

In addition to these primary industries, the Government petrochemical company diversified operations by producing salt, chlorine and acids, while metal pipe for the oil industry was also produced domestically from imported iron and steel. Innovation outside the oil industry was limited, although small local operators were producing cement, lime, sand-lime bricks, and lesser quantities of clay, stone, concrete, and other crude construction materials.

Table 3.3 World crude oil production, (1983 – 1991)

World Crude Oil Production, 1983 - 1991 (Million Barrels per Day)																			
		Selected OPEC Producers								Selected Non-OPEC Producers									
Year	Persian Gulf Nations	Iran	Iraq	Kuwait	Nigeria	Saudi Arabia	United Arab Emirates	Venezuela	Total OPEC	Canada	China	Mexico	Norway	Former U.S.S.R.	Russia	United Kingdom	United States	Total Non-OPEC	World
1983	11.08	2.44	1.01	1.06	1.24	5.09	1.15	1.80	16.57	1.36	2.12	2.69	.61	11.97	--	2.29	8.69	36.69	53.26
1984	10.78	2.17	1.21	1.16	1.39	4.66	1.15	1.80	16.50	1.44	2.30	2.78	.71	11.86	--	2.48	8.88	38.00	54.50
1985	9.63	2.25	1.43	1.02	1.50	3.39	1.19	1.68	15.37	1.47	2.51	2.75	.77	11.59	--	2.53	8.97	38.60	53.97
1986	11.70	2.04	1.69	1.42	1.47	4.87	1.33	1.79	17.46	1.47	2.62	2.44	.84	11.90	--	2.54	8.68	38.74	56.20
1987	12.10	2.30	2.08	1.59	1.34	4.27	1.54	1.75	17.71	1.54	2.69	2.55	.98	12.05	--	2.41	8.35	38.92	56.63
1988	13.46	2.24	2.69	1.49	1.45	5.09	1.57	1.90	19.74	1.62	2.73	2.51	1.11	12.05	--	2.23	8.14	38.96	58.69
1989	14.84	2.81	2.90	1.78	1.72	5.06	1.86	1.91	21.40	1.56	2.76	2.52	1.48	11.72	--	1.80	7.61	38.40	59.79
1990	15.28	3.09	2.04	1.18	1.81	6.41	2.12	2.14	22.49	1.55	2.77	2.55	1.63	10.98	--	1.82	7.36	38.00	60.49
1991	14.74	3.31	.31	.19	1.89	8.12	2.39	2.38	22.48	1.55	2.84	2.68	1.87	9.99	--	1.80	7.42	37.71	60.19

Source: Arab Petroleum Research Center, Arab Oil and Gas Bulletin (2009)

Table 3.4 Comparison of Government Oil revenue in relation to non Oil revenue

Year	Oil revenue	Non-oil revenue
1984/85	90.9%	9.1%
1985/86	89.3%	10.7%
1986/87	85.7%	14.3%
1987/88	88.4%	11.6%
1988/89	85.9%	14.1%
1989/90	90.8%	9.2%
1990/91	90.2%	9.8%
1991/92	76.6%	23.4%
1992/93	88.2%	11.8%
1993/94	83.8%	16.2%
1994/95	89.8%	10.2%

Source: Al-Hijji, Y.Y. (1997) Old Kuwait: Memories in Pictures. Almansouria, Kuwait: Center for Research and Studies on Kuwait.

3.4.6 From invasion and the decade of reconstruction (1992-today)

Restoration of the oil industry, which was heavily damaged during the Iraqi invasion, was a top priority given that oil revenues largely underpinned government spending programmes. Today the oil industry, producing an average of 2.098 million barrels per day (Table 3.5), has fully recovered from the attack of the Iraqi invasion, and the State-owned KPC is estimated to be the seventh largest oil company in the world. KPC has extensive overseas operations, including refineries and large downstream distribution networks in Western Europe, and exports almost 78 per cent of its products to Non-Arab States.

Findings provided by the Kuwait Institute of Banking (2002) suggest that Kuwaiti commercial banks currently enjoy a competitive position vis-à-vis the banks of the other GCC member states in terms of assets, number of branches, number of ATM machines, number of employees and number of subsidiaries. While these banks also enjoy a competitive advantage in terms of net profit attained (they rank second among the GCC banks) they performed less well in terms of return on assets (where they rank fourth). Moreover, the Kuwait Institute of Banking (2002) suggests that Kuwaiti banks' expenditure on staff training courses is low (they rank last) compared to other GCC banks. Therefore, Kuwaiti banks need to allocate further financial resources for staff development purposes. While Kuwaiti banks enjoy a relative competitive advantage in relation to expenditure on research and development (ranking first among GCC banks) they do not enjoy such an advantage in the area of expenditure on information technology and communications (where they rank fourth).

Table 3.5 World Crude Oil Production 1992 – 2008

World Crude Oil Production, 1960-2008 (Million Barrels per Day)																			
		Selected OPEC Producers								**Selected Non-OPEC Producers**									
Year	**Persian Gulf Nations**	**Iran**	**Iraq**	**Kuwait**	**Nigeria**	**Saudi Arabia**	**United Arab Emirates**	**Venezuela**	**Total OPEC**	**Canada**	**China**	**Mexico**	**Norway**	**Former U.S.S.R.**	**Russia**	**United Kingdom**	**United States**	**Total Non-OPEC**	**World**
1992	15.97	3.43	.43	1.06	1.94	8.33	2.27	2.37	R 23.74	1.61	2.85	2.67	2.13	– –	7.63	1.83	7.17	36.37	60.12
1993	16.71	3.54	.51	1.85	1.96	8.20	2.16	2.45	24.46	1.68	2.89	2.67	2.28	– –	6.73	1.92	6.85	35.71	60.17
1994	16.96	3.62	.55	2.03	1.93	8.12	2.19	2.59	24.90	1.75	2.94	2.69	2.57	– –	6.14	2.37	6.66	36.20	61.10
1995	17.21	3.64	.56	2.06	1.99	8.23	2.23	2.75	25.54	1.81	2.99	2.62	2.77	– –	6.00	2.49	6.56	36.85	62.38
1996	17.37	3.69	.58	2.06	2.00	8.22	2.28	2.94	26.02	1.84	3.13	2.86	3.09	– –	5.85	2.57	6.46	37.73	63.75
1997	18.10	3.66	1.16	2.01	2.13	8.36	2.32	3.28	27.29	1.92	3.20	3.02	3.14	– –	5.92	2.52	6.45	38.45	65.74
1998	19.34	3.63	2.15	2.09	2.15	8.39	2.35	3.17	28.37	1.98	3.20	3.07	3.01	– –	5.85	2.62	6.25	38.60	66.97
1999	18.67	3.56	2.51	1.90	2.13	7.83	2.17	2.83	27.22	1.91	3.20	2.91	3.02	– –	6.08	2.68	5.88	38.70	65.92
2000	19.89	3.70	2.57	2.08	2.17	8.40	2.37	3.16	28.98	1.98	3.25	3.01	3.22	– –	6.48	2.28	5.82	39.52	68.49
2001	19.10	3.72	2.39	2.00	2.26	8.03	2.21	3.01	28.16	2.03	3.30	3.13	3.23	– –	6.92	2.28	5.80	39.94	68.10
2002	17.79	3.44	2.02	1.89	2.12	7.63	2.08	2.60	26.39	2.17	3.39	3.18	3.13	– –	7.41	2.29	5.75	40.77	67.16
2003	19.06	3.74	1.31	2.14	2.28	8.78	2.35	2.34	27.98	2.31	3.41	3.37	3.04	– –	8.13	2.09	5.68	41.45	69.43
2004	20.79	4.00	2.01	2.38	2.33	9.10	2.48	2.56	30.41	2.40	3.49	3.38	2.95	– –	8.80	1.85	5.42	42.08	72.49
2005	21.50	4.14	1.88	2.53	2.63	9.55	2.54	2.56	31.87	2.37	3.61	3.33	2.70	– –	9.04	1.65	5.18	41.87	73.74
2006	21.23	4.03	2.00	2.54	2.44	9.15	2.64	2.51	31.59	2.53	3.67	3.26	2.49	– –	9.25	1.49	5.10	41.87	73.46
2007	20.67	3.91	2.09	2.46	2.35	8.72	2.60	2.43	31.21	2.62	3.73	3.08	2.27	– –	9.44	1.50	5.06	41.80	73.01
2008	21.87	4.03	2.36	2.59	2.17	9.26	2.68	2.39	32.47	2.59	3.79	2.79	2.18	– –	9.36	1.39	4.96	41.31	73.78

Source: Arab Petroleum Research Center, Arab Oil and Gas Bulletin (2009)

Since its liberation from Iraqi occupation in February 1991, Kuwait has focused and continues to build on its pre-invasion oil-based economy. Education is a primary tool in achieving this. Institutions offering a western education curriculum have grown over the years, while the national university of the country (Kuwait University) has expanded and grown to be recognised internationally as one of the most prominent universities in the Gulf.

3.5 Kuwait seeking growth and development in the new millennium

Economic policies are required for national economic growth, yet economic growth cannot be separated from the cultural, ecological, social, political and spiritual environment in which it takes place. Economic growth and development cannot be seen as separate initiatives. Economic development drives social development and is, in turn, linked to peace, freedom, social stability and security. The United Nations Department of Economic and Social Affairs (2008) suggest that the promotion of social development requires an orientation of values, objectives and priorities towards the well-being of all and the strengthening and promotion of institutions and policies.

As oil revenues continue to flow, part of this blessing should be directed to education as this can be viewed as an engine of growth and innovation. Empirical evidence gathered by Gadir Ali (2009) suggests that strong correlations exist between education and social and economic indicators. Gadir Ali (2009) thus contends that further investment in human capital is crucial. Education, particularly of females, will result in the empowerment of females in societies where they have (in the past) been accorded fewer educational opportunities than males. To strengthen education, an economy has to devote more resources to increased provision of better education at all levels, and the blessing of oil gives Kuwait a unique advantage in this area.

Creating a culture of savings ought to be considered a further integral element of growth. Research findings suggest that saving and investment are conducive to long-run economic growth and social development. The United Nations' Commission on Sustainable Development (2008) suggests that as countries become economically competitive and financially affluent, there is a danger in placing rapacious consumerism above savings and thus compromising the welfare of future generations. To this end the Kuwait Investment Authority is also a key player in promoting the savings objectives of Kuwait. There is also a need to encourage increased domestic and foreign investment, and to create and foster an entrepreneurial environment in which investment projects are motivated by profit incentives rather than by political reasons. One development in this field has been the extension of the operations of Kuwait Foreign Investment Bureau which encourages foreign investment in Kuwait.

Openness to the rest of the world is important for growth. Closed economies can no longer survive the onslaught of globalization. Imports and exports constitute a major proportion of an economy's balance of payments and regulations that are ineffective may render economic policies useless.

There is also an acute need to find ways to avoid an excessive reliance on the public sector in areas of production and provision of services. Historically, oil revenues have supported the growth of the public sector, yet private enterprise has proved to be a powerful engine of economic growth around the world.

The recent overall performance of the national economy, through a strengthening of economic activity rates in both the oil and non-oil sectors, has been encouraging. While many resource-based economies have suffered from profligate or corrupt governments, Kuwait has largely avoided this scenario, as detailed in Table 3.6 which gives an indication of the corruption level of Kuwait relative to other countries. Kuwait ranks 116th out of the 163 nations evaluated by the Transparency International index, and is seen as a good place to do business by the Economist, an indication of the responsible manner in which the government is addressing the resource "curse".

Table 3.6 Oil Importing Countries and Corruption

Country	Rank in terms of U.S crude oil imports	Oil barrels/day exported to the U.S – June 2006	Corruption index rank, by Transparency International (1 – most and 163 least)	Economist Global Business Environment grade
Canada	1	1,799	150	Very good
Mexico	2	1,734	86	Good
Saudi Arabia	3	1,549	86	Moderate
Venezuela	4	1,008	23	Very poor
Nigeria	5	996	14	Very poor
Iraq	6	617	2	**
Angola	7	525	14	Very poor
Algeria	8	474	75	Poor
Ecuador	9	282	23	Poor
Russia	10	216	35	Moderate
Colombia	11	211	105	Moderate
KUWAIT	**12**	**201**	**116**	**Good**
United Kingdom	13	185	151	Very Good
Equatorial Guinea	14	114	9	**
Libya	15	110	54	Very poor

Source: Transparency International - http://www.transparency.org/

The role of the private sector cannot be underestimated in this regard. Kuwait boasts of several major sectors. The Kuwait Stock Market shows the daily performance of the listed companies. The banking sector is a major sector that plays a vital role in innovation. Kuwaiti Banks are major players in the banking sector who earned several accolades and is highly rated internationally. There are fifty two listed investment companies on The Kuwait Stock Market. Over the last decade Kuwait witnessed an increase in investment in real estate and construction in comparison to the sudden drop in Dubai. Kuwait Real Estate Companies have been dealing in various real estate activities including the sale and purchase, leasing and renting of lands and buildings, development and construction of buildings. Moreover, the companies have also invested in shares or projects similar to its own activities. The insurance sector boasts several listed companies. Their activities include conventional insurance and Islamic financing. Moreover, the growing industrial sector also boasts several listed companies. As industrial trade barriers in Kuwait are being reduced, the result is that goods and services, technology, capital, labor and information are now moving more freely into Kuwait. Kuwait is preparing its industries to embark on an active program of industrial development. These industries provide a sound and sustainable base for Kuwait's development. This sector of the Kuwaiti economy is expanding both domestically and in the region. Private transportation is somewhat exclusively a small business but public transport is the State's prerogative and there are talks of privatization of the National flag carrier, namely Kuwait Airways.

3.6 Institutions supporting innovation in Kuwait

Four important institutions that support different forms of innovation and hopefully will help to build sustainable innovation capacity in Kuwait can be identified. Although these institutions are autonomous, they work collectively in promoting innovation at different levels. Moreover, it would not be uncommon to see a board member or political appointee serving as a member on one or more of the institutions. This is consistent with the approach adopted by Hong Kong. According to Yam (2011) when Britain handed the sovereignty of Hong Kong over to China in 1997, The Hong Kong Special Administration Region (HKSAR) government revisited its technology policies and took a number of measures to facilitate technology transfer and foster a culture of innovation in the community – this included the decision that it was important for the different institutions active in the technology field to collaborate (with members serving on more than one board a way of supporting this).

These four institutions are:

The Kuwait Investment Authority (KIA);
The Arab Fund for Social and Economic Development;
The Kuwait Foundation for the Advancement of Sciences (KFAS); and
Kuwait Institute for Scientific Research (KISR)

Firstly, **The Kuwait Investment Authority (KIA)** is an independent government organization entrusted with the management and administration of the country's sovereign wealth funds and for several reasons is considered as an organization that supports and promotes innovative thoughts through its financing initiatives. .

KIA sets benchmarks for educational and training standard in investment and securities analysis and works closely with universities. KIA offers ten prestigious academic scholarships per year to Kuwaiti Nationals seeking to obtain their MBA (Masters in Business Administration) degree from top ranked universities in the world with high academic standards. To ensure the integrity of the scholarship, the program is supervised by the Cultural Division in the Kuwait Embassies. Moreover, the personnel that are associated with KIA are key people from the private, public and no governmental sectors.The KIA's board of directors is headed by the Minister of Finance, with other seats allocated to the Energy Minister, Governor of the Central Bank of Kuwait, Undersecretary of the Ministry of Finance and five other nationals who are experts in the field.KIA is the custodian of all funds and is ultimately responsible for funding of projects that result in the growth and development of Kuwait.

The two specific funds which represent the principal holdings currently under the management of the KIA are as follows:

The General Reserve Fund (GRF) – the fund for receiving all revenues from which state budgetary expenditures are paid; and

The Future Generations Fund (FGF) – created in 1976 by transferring 50% from the GRF at that time. In addition, 10% of all state revenues are transferred to the FGF on an annual basis and all investment income is reinvested. In particular, global investments are focused on private equity, real estate and hedge funds within the agreed strategic asset allocation.

The organization might also be tasked with managing and administering other funds entrusted to it by the Minister of Finance and these funds, in addition to the GRF and FGF, are likely to be invested in the local, regional and international markets. KIA is estimated to hold in excess of $200 billion of assets, and is reportedly one of the largest sovereign wealth funds in the world. Some of the principal KIA sovereign wealth enterprises are the following:

National Technology Enterprises Company – The National Technology Enterprises Company (NTEC) was established, in November of 2002 by the Kuwait Council of Ministers with US$311million in start-up capital and is a fully owned subsidiary of the Kuwait Investment Authority. NTEC is mandated by the Kuwait Council of

Ministers to play help servicing the major industrial and commercial stakeholders in Kuwait with their technology needs. NTEC's business model is that of a Technology Projects Development company utilizing investment tools such as Private Equity, Venture Capital and Direct Investments to initiate and stimulate Technology Projects in Kuwait and the local region;

St Martin's Property – the real estate investment arm of the KIA which has held significant holdings in the City of London for over eighty years;

Kuwait Real Estate Investment Consortium – of which KIA owns 99.127%. The consortium carries out real estate transactions, investment in securities and investment in portfolio management activities domestically and internationally; and Kuwait China Investment Company – 15% stake owned by the KIA and it represents the key investment vehicle for the organization in Asia dealing with sectors as diverse as agriculture, energy and financial services.

In addition to its main headquarters in Kuwait city, KIA also has a branch office (the Kuwait Investment Office) in the City of London which was established in 1953 with the aim of investing surplus oil revenues and thereby reducing dependency on the country's finite oil resources. Its principal objective is to manage that portion of the FGF allocated to it and to achieve a fair return over the long term by acting as a global investor covering the areas of equities, fixed income, treasury, private equity and property. Indeed, the mission statement of the KIA underpins this requirement of achieving long term investment returns on the financial reserves of the State of Kuwait and thereby providing the country with an alternative to oil revenues. These goals are embodied in the principal legislation governing the activities of the KIA – Law 47/1982.

Moreover, KIA established a Training Department in 1995. A training plan was developed to achieve the following goals:

- Promote staff performance and seek to enhance and develop career and technical skills.
- Contribute to national development by investing in human resources.

In order to achieve these goals, KIA has set up a recognized and distinguished training centre equipped with the most innovative instruction and demonstration technologies. The Training Department focuses, among others, on the following:

- Improving the performance of KIA staff;
- Training new Kuwaiti Graduates;
- Training the staff of local financial institutions and companies;
- The MBA scholarships program;
- Kuwait Automotive Academy; and
- World Bank Recruitment Program.

KIA strongly believes that knowledge (human capital) adds critically to the performance of the private sector.

Secondly, **The Arab Fund for Social and Economic Development**, based in the State of Kuwait and established in 1974, is a regional financial institution overseen by the Central Bank of Kuwait, whose principal aim is the funding of economic and social development in the Arab world through the financing of public and private investment projects and by providing grants and expertise.

In the fulfillment of this aim, the loans extended to member countries of the organization are offered on concessionary terms by providing easy financing conditions to assist these member countries implement their development plans. In the conduct of its operations, the Fund follows guidelines on neutrality and organizes itself on the basis of substantive rules to ensure independence from any political considerations.

The Fund has recently focused much of its attention on the funding of private sector development projects in the Arab countries and its activities have been directed at providing direct loans, equity investment, guarantees, lines of credit, as well as institutional support and advisory services. Projects deemed vital to the Arab world and those assuming high priority in national development plans are given precedence. Before any project receives support from the Fund, it is subject to a thorough evaluation of its technical and economic feasibility during which sound legal, organizational and technical structures are created to ensure the success of the project.

When assessing the fairest distribution of its financial loans, the Fund gives priority to inter-Arab projects which are likely to promote cooperation, integration and solidarity amongst member countries. Hence the emphasis on contributing to projects involving the interconnection and interdependence of Arab countries such as electrical power, transportation and communications. In the field of social development and the reduction of poverty, the Fund's financial projects cover areas such as health care, education, drinking water, rural development and social welfare. In its support for the least developed countries in the Arab world, the Fund has offered unstinting help to the Palestinian people and has provided grants to educational institutions, universities and professional and social associations. The principal loan beneficiaries at the end of 2009 were Yemen (89 loans), Morocco (59) and Syria (49), with loans totaling, in the case of Yemen, 639,700 Kuwaiti Dinars. In its endeavours, the Fund encourages participation from the member states in the assessment and development of the projects and seeks to gain from their existing expertise in project evaluation, implementation and supervision.

The main activity of the Fund centres round providing loans, on concessionary terms, to governments and public corporations and enterprises of member states. In this context interest rates are reduced to 2.5% for low-income Arab countries and 3% for others. Furthermore, the Fund has increased the grace period before loan repayments start to between four and six years and extended the loan term to between 22 and 25 years. At the same time, the Fund concentrates significant effort on the development of the private sector by financing operations beneficial both to the developmental role of the Fund and the private sector itself. To this end, a Special Account was set up by the Fund in 2010 – with members pledging $1.2 billion – to assist small and medium size private sector projects in member States. In addition, approximately 5% of the Fund's operational profits are directed towards providing grants aimed at providing institutional support to member states, improving efficiency in project implementation and conducting technical and economic studies in key developmental areas.

Furthermore, in addition to providing loans and grants to finance development projects, the Fund also provides a limited number of scholarships to Arab PhD holders in different specialist fields. The Fellowships Programme was launched in December 1997 to enable selected PhD holders to conduct advanced research and/or lecture in some of the world's most prestigious universities. Since its inception, the programme has awarded about 66 scholarships to Arab researchers across a variety of specializations. From an innovation perspective direct recipient of funds are encouraged to further their scholarly work in Kuwait.

Thirdly, **the Kuwait Foundation for the Advancement of Sciences (KFAS)** is a private, non-profit making organization established by Emiri decree in 1976 and is managed and administered by a Board of Directors (BOD), chaired by H.H. the Amir of the State of Kuwait. Financial contributions to KFAS (1% of the annual net profit) are received from Kuwait shareholding companies, which form the main source of funding for the organization. Most recent figures from KFAS show that 259 investment companies, 135 service sector companies and 125 real estate companies support the activities of the Foundation.

The principal goal of the organization is to promote scientific and technological advancement within the country and the region as a whole by supporting scientific projects, the scientific community and the country's scientific infrastructure. The foundation seeks to achieve this objective by:

Providing financial assistance for the conduct of research projects in both the basic and applied sciences;
Supporting projects that are considered to be of national priority and strengthening indigenous scientific talents and initiatives;
Awarding honours and bestowing recognition on outstanding scientific achievements at national, regional and international levels;
Publishing scientific journals, books and encyclopedias and thereby provide a reference library of Arabic scientific resources;
Developing the resources and skills of Kuwait nationals through scholarships, training, workshops, seminars and conferences; and
Promoting scientific and cultural awareness.

In terms of financial assistance to research projects, KFAS has funded 543 such projects in the areas of biological sciences, engineering sciences, technology, medicine, natural sciences, social sciences and humanities. Its 2011-2012 budget is estimated to be aroundUS$4 billion. With regard to funding for cultural activities, the organization focuses its resources on raising awareness of the role of Islamic and Arab civilization throughout the world and on enhancing mutual cooperation between the scientific, social and cultural organizations in Kuwait with their counterparts abroad. To achieve the latter objective, resources are allocated to the publication of scientific books and magazines, the holding of scientific symposia and the running of competitions. Indeed, the foundation makes large-scale contributions to the field of publishing and has published a substantial collection of scientific books as well as producing films and documentaries recording the scientific achievements of both Arab and foreign scientists. Other activities in the area of raising scientific awareness include supporting student participation in international scientific Olympiads, the production of documentary films and the organization of cultural events.

To further enhance and promote scientific research, KFAS awards outstanding local, regional and international scientists and researchers the following, scientific achievement awards:

The Kuwait Award which is awarded every year for the promotion of science and culture across a variety of branches;

The awards for outstanding work carried out by authors and translators and granted during Kuwait's annual book fair; and

Awards to encourage the work of distinguished university and college graduates.

In addition, specific achievement awards are presented to outstanding contributions to medical science, agriculture, animal and fish resources, the best research papers and studies into the repercussions of the Iraqi invasion of Kuwait. Some examples of the financial support given to the country's national scientific infrastructure are the $1bn allocated for the establishment of the Scientific Centre and the establishment of the Dasman Centre for Research and Treatment of Diabetes.

The Foundation is managed by a Board of Directors, chaired by H.H. the Emir, and comprises six members selected by Kuwait shareholding companies for a period of three years. The Board of Directors appoints a Director General who manages the scientific, administrative and financial activities of KFAS and monitors the implementation of the policies set by the Board.

Fourthly, **the Kuwait Institute for Scientific Research (KISR)** was set up in 1967 to conduct applied scientific research in three areas: petroleum, desert agriculture and marine biology. This applied scientific research was meant to focus on domestic industry, energy, agriculture and the national economy in order to make a contribution to the social and economic development of the state. In 1981 KISR was legally designated as an independent public institution governed by a Board of Trustees chaired by a minister from the Council of Ministers. The institute was also tasked with carrying out studies relating to the preservation of the environment, resources of natural wealth and their discovery, sources of water and energy and the methods of agricultural exploitation.

In fulfillment of this mission, KISR has defined its objectives as follows:

Conduct scientific research to benefit national industry and the environment;
Explore the country's natural resources for their exploitation and preservation;
Provide scientific and technological consultation services to the government;
Foster relations and conduct mutual research with higher education institutes in Kuwait and overseas; and
Contribute to the economic and social development goals of the State.

KISR classifies its main research activities into five principal areas each with their own mission statement and specific goals. They are as follows:

Environment and Urban Development dealing with pollutants in the terrestrial ecosystems, industrial and domestic waste and accompanying safety measures;
Food Resources and Marine Sciences dealing with animal husbandry, plant production, soil sciences and native plant cultivation and wildlife preservation;
Petroleum Research and Studies Centre set up in 2000 to focus on petroleum production, refining, polymer and petrochemicals, materials corrosion and health and safety;
Techno Economics dealing with strategic economic issues for the future prosperity of the State, such as privatization, globalization, the development of business plans and market analysis for prospective investment opportunities as well as developing decision support tools that will optimize resource allocation in the service and industrial sectors and other activities that include the development of web applications for mathematical and economic models and the conduct of feasibility studies, statistical analysis, experimental design and mathematical modeling for research projects at KISR; and
Water Resources dealing with the management and optimal utilization of available resources, reducing the costs of production, desalination and treated wastewater. This area of research also deals with the protection of water resources from environmental hazards and looking at ways of public/private sector partnerships to increase efficiency and productivity.

Analytical support and services for KISR is provided by the Central Analytical Laboratory with its five specialist laboratories covering wet chemistry, chromatography, spectroscopy, physical chemistry and trace metals and the Building Materials Lab for the physical and mechanical testing of cement, masonry components and general construction materials. In addition to the lab facilities, KISR also has specialized centres – the National Seismic Network monitoring seismic activity in Kuwait, the National Meteorological Network with its nine meteorological stations throughout the country and the Remote Sensing Lab (Satellite Imaging) for the production, among other things of maps, data and analyses of geo-sphere, marine environments, atmosphere, natural hazards and the dynamic processes of the planet.

KISR continues to host a number of scientific conferences covering subjects as diverse as disaster management, water resources and the environmental challenges, enhanced building operations and the impact of climate change on the agriculture and biodiversity of the Arab region. In addition, the institute publishes a whole range of books, articles and scientific reports as well as its Science and Technology magazine, and conducts student courses, workshops and seminars both for in-house staff and for the general public to increase understanding and awareness of major scientific, environmental and biodiversity issues. These institutions among many others play pivotal role in policy issues relating to innovation. Table 3.7 presents the percentage of the budget on different activities of KISR (2011-2012) as reflected on its website.

Gauging from the data presented, KISR's approach towards innovation, is somewhat different from the west. In Kuwait an all-encompassing approach is developed. Diffusion of innovation is vitally important.

Table 3.7 2011-12 Annual Budget for independent Project

Elaboration of the most appropriate networking institutional mechanism through consultation with member states	3.448%
Inception workshop on basic requirements and resources for the development of sustainable regional monitoring system for POPs in ambient air.	10.344%
Technical regional Expert Group Meeting for the Network partner to discuss means and options of for the development of a regional ambient air POPs monitoring system.	10.344%
Preparation of a project proposal on "Air monitoring of POPs in West Asia region" (to be submitted to GEF and other funding agencies).	6.896%
Ongoing pilot project at KISR for monitoring POPs in ambient air in Kuwait.	34.482%
Communications with UNIDO and other partners to establish a regional forum to include all West Asia Parties.	12.073%
Collation of information related to POPs issues (guidelines, technical articles, reports, web links, etc).	3.448%
Facilitation of the potential development of a database on POPs in the region (to be hosted at KISR).	1.724%
Development of a pilot web site with access interface.	20.689%
The 4th UNIDO Consultation Meeting for BAT/BEP Forum for the GCC Region.	6.896%

3.7 Conclusion

This chapter documents Kuwait economic, political and social history in relation to the growth and development of the oil and oil related industries. The intention is to highlight the rather difficult task that faces a benevolent ruling government benefiting from high levels of natural resource-based revenues. Is there a relationship between natural resource wealth, the design of democratic institutions and optimal collective decisions? Optimum decision making is defined as achieving goals important to the developmental process such as deep and equitable reforms placing the government in a central position in caring for its citizens. The Kuwaiti constitution is designed to distribute power among actors, generate efficiency in governance, and govern the interactions of actors. At the same time to move forward, it is imperative to reduce reliance upon oil and oil-backed resources. To this end innovation and the development of non-oil sectors must be a non-negotiable task for future governments. To begin to address this, the next chapter considers policies and practices associated with these issues. Whilst this chapter has provided a brief account of economic changes during significant periods showing the country has been modernised, its people's links with their heritage are as strong as ever. Kuwait's wealth -- Black Gold -- may be buried underground, but its spirit is still vibrantly free and alive and will continue to influence developmental strategies in the years to come. The challenge is to ensure that the Black Gold benefits - and not curses - future generations of Kuwaitis.

CHAPTER FOUR

LEVERAGING INNOVATION AND INCREASING RETURN ON INNOVATION FOR LONG TERM SECURITY

"Economic life is a struggle between those that want to change things, and those that want to keep things as they are. (As in any war) the advantage lies with those that attack. Technological innovation is a choice weapon in this struggle. Innovation is a strategy for survival."
(Foster, 1986)

4.1 Introduction

The preceding chapters noted that income accruing as a result of the discovery of oil in Kuwait rapidly changed Kuwait's economic priorities, bringing new opportunities and at the same time new challenges. It has also been highlighted that the government of Kuwait is concerned about many important issues relating to the depletion of non-renewable resources. One of the unique challenges facing Kuwait is what collective action is necessary to safeguard time honoured traditions that combine economic prosperity with solidarity. Today Kuwait is in need of new commitments on the part of its citizens and decisive actions in political leadership. Instead of maintaining structures and organizations that have shown themselves unable to deal with the challenges that face Kuwait, Kuwait must be ready to support structural changes. This in particular requires a prioritization of resources towards education, research and development. Kuwait can only become comprehensively innovative if all sectors support the development of innovative products and services. The need for innovation should be identified by all sectors. Increasingly the various different sectors appear as focal points for learning and knowledge creation in this new age of global, knowledge-intensive capitalism and, when any one sector would get to look at another as an important source of innovation and economic growth, this form of innovation diffusion among sectors becomes important to foster a reciprocal understanding and mutual confidence among the various different sectors. Strategic issues entailing innovation require the involvement of all parties. These include businesses, the public sector, producers and consumers. A wide-ranging partnership for innovation is necessary, particularly when a country's resources are in question. To establish an optimal framework and develop potential for innovation, the prospect of an innovation-friendly market must be widely accepted.

In light of these issues this chapter critically examines innovation as a factor associated with wealth creation in Kuwait and examines the country's economic priorities. It explores the decisions a country must make when faced with an abundance of a natural resource and asymmetric problems such as limited access to technical knowledge, a failure to engage - particularly in long-term planning, weak institutional structures, and missing mechanisms of accountability. This chapter identifies innovation as a solution to the resource curse and examines what institutional innovations are necessary to align the incentives of key domestic and international players to facilitate innovation. It then lays out a path for improving the management of the wealth of Kuwait through innovation to meet the challenge of the resource curse.

4.2 National systems of innovation – EU Survey of Innovation

In its response to increased competition and globalisation the European Council argued for increased and enhanced efforts to improve the Union's performance in innovation. In March 2000 in the picturesque city of Lisbon the Union set itself the goal of becoming the most competitive and dynamic knowledge-based economy in the world within the next decade. Fine words one may say, but precisely how does one set about achieving this laudable goal? A strategy was developed and presented in Stockholm in March 2001. The strategy was to build on the economic convergence that had been developed over the past 10 years within the EU single market and to coordinate an 'open method' of developing policies for creating new skills, knowledge and innovation. To support this approach the European Commission stated that there was a need for an assessment of how member countries were performing in the area of innovation. The idea of a 'Scoreboard' was launched to indicate the performance of member states. This would be conducted every year as a way of assessing the performance of member countries. It is essentially a benchmarking exercise where the European Union can assess its performance against other countries, most notably Japan and the United States.

The 2005 Scoreboard also analyses Bulgaria, Romania, Turkey, Iceland and Switzerland. This is an extremely ambitious project to try to assess innovative ability. There have been many studies over the past two decades that have tried to identify the factors necessary for innovation to occur (see Table 4.1), and while many factors have been identified many of these are necessary but not sufficient in themselves. The latest European Innovation Survey was conducted in 2009 (EIS, 2010). Figure 4.1, shows a map of the European Union indicating five performance groups, ranging from the highest to the lowest overall performers. Table 4.1 illustrates the how the EIS (2010) divides the countries into four key groups.

Moreover, some governments have attempted to develop 'innovation toolkits' and 'scorecards' to try to help firms in their own countries to become more innovative (UK Department of Trade and Industry). Most of these have not been successful. This ambitious project by the European Union is full of limitations and is generally regarded as over-simplistic. This is largely because the economic conditions of the member countries are so very different and all have a wide variety of strengths and weaknesses. None the less, in order to assess where the European Union should target help and the precise type of help required by each member it is necessary to analyse the innovative performance of countries.

Since innovation is the result of numerous interactions between key organizations and institutions, the national innovation system suggests that the flows of technology and information among people, enterprises and institutions are vital to any innovative process. The Organisation for Economic Co-operation and Development (OECD) identify innovation and technology development as the result of a complex set of relationships among actors in the system, which includes enterprises, universities and government research institutes. The Organisation goes further and suggests that for policy-makers, an understanding of the national innovation system can help identify leverage points for enhancing innovative performance and overall competitiveness. It can assist in identifying dissonance within the system, both among institutions and in relation to government policies, which can prevent technology development and innovation. With increasing international competition, the time to market products is becoming shorter. Therefore a collaborative innovation strategy through the promotion of efficient networks of suppliers, knowledge institutes, intermediaries, customers, users, competitors need to be established. The idea that is advocated is that a decision to innovate needs the collaborative decision-making of several stakeholders that make up the national innovation system of the country.

Table 4.1 Innovation Growth Leaders

GROUP	GROWTH RATE	GROWTH LEADERS	MODERATE GROWERS	SLOW GROWERS
INNOVATION LEADERS	1.5%	Switzerland (CH)	Finland (FI), Germany (DE)	Denmark (DK), Sweden (SE), United Kingdom (UK).
INNOVATION FOLLOWERS	2.7%	Cyprus (CY), Estonia (EE)	Iceland (IS), Slovenia (SI),	Austria (AT), Belgium (BE), France (FR), Ireland (IE), Luxembourg (LU), Netherlands (NL).
MODERATE INNOVATORS	3.3%	Czech Republic (CZ), Greece (GR), Malta (MT), Portugal (PT)	Hungary (HU), Lithuania (LT), Poland (PL), Slovakia (SK),	Italy (IT), Norway (NO), Spain (ES).
CATCHING-UP COUNTRIES	5.5%	Bulgaria (BG), Romania (RO)	Latvia (LV),	Turkey (TR), Croatia (HR).

Goetheer and Heijs (2011) in their study - key players in national systems of innovation - identify the importance of collaboration in networks for the diffusion of knowledge and the development of innovations. However, they contend that the measurements to determine key players in systems of innovation are scarce. They build on the social network theory, to assess methodological approaches to determine key players in national systems of innovation. In this regard Goetheer and Heijs (2011) develop and assess three approaches, namely, top down, bottom up and systems approach to determine which approach gives more representative results. Based on their research, Goetheer and Heijs (2011) recommend the systems approach as it reveals more representative results.

In pursuit of the EU vision, Arab States realized that the concept of Information Society is not restricted to a set of IT applications, but also includes a broader and more extensive vision that gives society the chance to make more progress and achieve an effective participation in the 21st century civilization. The World Summit on the Information Society (WSIS) was held in two phases (http://www.itu.int/wsis/index.html). The **first phase** took place in Geneva hosted by the Government of Switzerland from **10 to 12 December 2003**, and the **second phase** took place in Tunis hosted by the Government of Tunisia, from **16 to 18 November 2005.** It is indeed historic that the occasion took place in Tunis reflecting the importance of ICT not only in the developed western world but also in Africa and the Arab world. According to The Economic and Social Commission for Western Asia (ESCWA, 2007) the all encompassing EU vision can be achieved through the development and modernization of ICT infrastructure, the increased reach of Internet as an access tool to information and knowledge, the building of trust and security in IT usage, and the capacity building of individuals to effectively participate in information society by increasing cultural and IT awareness, developing education, encouraging the use of IT in general, and electronic transactions in particular, enhancing the role of the government, the business sector, and civil society organizations in using ICT for the development of all its citizens. The setting of relevant legislations related to electronic transaction becomes vital within the cooperation and implementation framework of international agreements and conventions.

4.3 Improving the innovation performance of the EU

Figure 4.1 shows a map of the European Union indicating five performance groups, ranging from the highest to the lowest overall performers. All the elements in the Scoreboard are necessary but not sufficient in themselves to ensure that innovation occurs. For example, in this chapter we have seen the example of Turkey, a late-industrialising country on the edge of Europe, a country with a population of 60 million, already a member of the North Atlantic Treaty Organisation (NATO) and a prospective member of the European Union. Turkey is a good example of a late-industrialising economy. Sitting on the edge of Europe and bestriding two continents, Turkey should be in a position to develop a successful economy. However, in Turkey there seems to be a missing link in terms of the innovative intention and capabilities of enterprises. Turkey needs to put in place many of the things detailed in the Scoreboard. This would surely help to develop enterprise in the country, but it will not convert Turkey into a Germany or Finland overnight.

Figure 4.1 Map of the European Union indication five performance groups

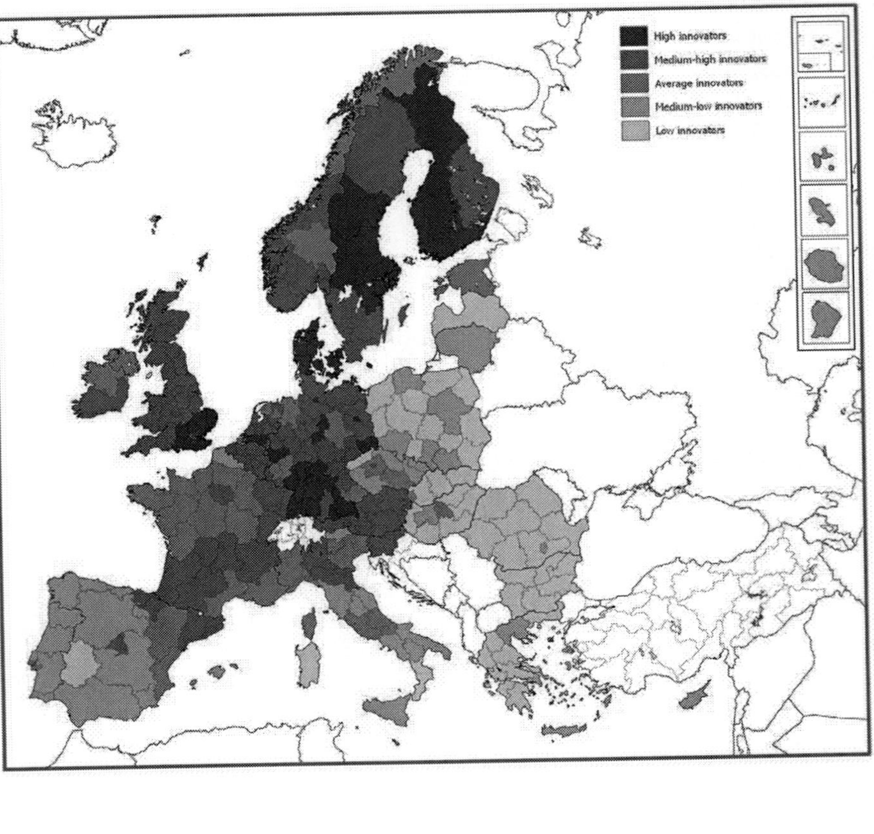

By identifying, comparing and disseminating best practices in financing and technology transfer, Europe can improve its innovation performance. One area that needs particular attention is the overall perception of the entrepreneur. The image of the entrepreneur needs to have greater value, as in the United States where the drive to try to market new products, with the in-built risk of failure, is seen much more positively than in Europe.

The Scoreboard may be helpful to governmental policy makers in deciding where to invest substantial sums of money. However, the first two chapters of this book have emphasised that firms behave differently given similar circumstances and that some firms appeared to be more successful than others. Given this, the Scoreboard's practical help is likely to be extremely limited.

4.4 National innovation systems in Kuwait

The different innovation capabilities of individual firms are regarded as the key components of a country's innovation system (Yama, 2011). At the outset it must be stated that Kuwait was visibly represented by senior government officials at the first world summit in Geneva in 2003 and declared it commitment to the declaration: "We, the representatives of the peoples of the world, assembled in Geneva from 10-12 December 2003 for the first phase of the World Summit on the Information Society, declare our common desire and commitment to build a people-centered, inclusive and development-oriented Information Society, where everyone can create, access, utilize and share information and knowledge, enabling individuals, communities and peoples to achieve their full potential in promoting their sustainable development and improving their quality of life, premised on the purposes and principles of the Charter of the United Nations and respecting fully and upholding the Universal Declaration of Human Rights". Consistent with this declaration a fundamental hypothesis of the national innovation systems approach stresses that the **flows of technology and information** among stakeholders. The implementation of innovation and technology developments are the result of a complex set of relationships among a host of individuals. A national innovation system is most valuable in Kuwait's context. The Economic and Social Commission for Western Asia (ESCWA, 2007) of the United Nations conducted the study: "National profile of the information society in Kuwait" - offer a summary of the achievements and progress made towards building a comprehensive information society that transforms ICT into an effective tool for sustainable development. Moreover, Kuwait Central Agency for IT was established in 2006 and its activities are reflected on its portal;

http://www.cait.gov.kw. This reflects the determination of the State of Kuwait for a National Strategy to build an information society to contribute to the growth of Kuwaitis and the use of IT as a development tool. This strategy follows the action plan of the World Summit on the Information Society (WSIS) and the regional action plan of West Asia countries. The Central Agency for Information Technology in Kuwait runs awareness and training media campaign to establish e-culture at all social levels, and to encourage the use of IT in all aspects of life, and thus contributes to the development of a comprehensive information society. The public/private partnership (ppp) or multi-sector partnership (msp) in Kuwait are reflected the private sector's active role in it development consistent with the regulations of government and the Central Agency for Information Technology. Kuwait Central Agency for IT states that the Non-governmental organizations have taken part in a number of committees and task forces established by governmental institutions in the ICT field in relation with the e-government project and e-services, in order to offer counsel and participate in their implementation and raise awareness of them. This approach is critical particularly in Kuwait since time honoured traditions in Kuwait can in no way be dismissed insensitively.

4.5 Progress of Kuwait towards fulfillment of national policies and strategies

The Kuwaiti government supports the application of e-government as a first step towards the achievement of e-Kuwait. This support is vitally important since it immediately creates a mindset of change. Eight governmental bodies provide electronic information services on individuals or corporations, in addition to a number of electronic transactions. Moreover, the necessary draft laws pertaining to electronic services and transactions are already in place. The number of electronic services provided by private sector companies has increased, and all Kuwaiti banks offer e-services, and electronic trading has become one of the main activities of the Kuwait Stock Exchange (KSE) and it is not surprising to find aged investors on the KSE using technology for their trading activities. There are electronic sale, purchasing, and payment transactions that have emerged either via the Internet or SMS.

The ICT infrastructure (both wired and wireless) covers the entire Kuwaiti territory through the Ministry of Telecommunications, mobile phone companies, and Internet service providers (ISPs). The Ministry of Telecommunications manages and supervises the wired network in

Kuwait, which has seen a growth in the number of switchboards, an increase in capacity, and a rise in the demand for all types of services. In 2008, the Ministry increased telephone numbers from 7 to 8 digits in order to accommodate the demand on landlines and make numbers more in line with international standards. Mobile phone companies and ISPs rely on the Ministry of Telecommunications infrastructure to operate their services, since they lease electronic circuits and sites that belong to the Ministry. There are four main ISPs in Kuwait that offer equally competitive services and Wi-Fi access points have increased in public places such as malls, service centers, and cafes, thus leading to the decrease in the number of Internet cafes. Moreover, Wi-Fi is widely used in households withDSL and is extremely popular with many families. This is mostly due to incentives and promotions ISPs offer to subscribers. There are currently three mobile phone companies in Kuwait. These companies have been transferring data and using the Internet through their wireless networks that cover the entire Kuwaiti territory. The speed at which Kuwait is progressing is remarkable. This is not surprising since Kuwait is a relatively small country and at the same time innovation strategy adopted by Kuwait is consistent with the rest of the world but with much more sensitivity ensuring that time honoured traditions are not violated in any way.

Kuwait recognizes with the rest of the world that building an inclusive Information Society requires new forms of solidarity, partnership and cooperation among governments and other stakeholders, namely, the private sector, civil society and international organizations and Kuwait realizes the ambitious goal of this declaration as espoused by the 2003 World Summit on the Information Society – that Kuwait will bridge the digital divide and ensuring harmonious, fair and equitable development for all - will require strong commitment by all stakeholders, we call for digital solidarity, both at national and international levels.

4.6 Socio-economic imperatives of Kuwait

Any thoughts associated with r*esource curse reduction* should ideally begin with an agenda for improving the management of resources that generate wealth. According to Al-Hassan Delgado and Omran (2007), setting priorities and economic initiatives are imperative in order to overcome the curse. However, this needs to be done within the confines of the socio-political paradigm (as we have outlined in the preceding chapter). In Kuwait in particular, as it is in developing economies in general, there ought to be a 'buy-in' to consumer trust and confidence in innovation (Al-Melhem, 1998). Blas and Kerr (2007), examining soaring consumer spending in the Gulf Countries, contend that the confidence of consumers in novel products and services is facilitated by a robust system of production and consumption protection. Porter (1999) and Delgado, Porter and Stern (2008) suggest that in markets where consumer confidence is high, the introduction of new products and innovative products is easier.

In this new economic order, Al-Melhelm (1998) contends that Kuwait cannot thrive unless it becomes more inventive, responds more effectively to market preferences and increases innovation. A literature survey of some poorer countries, examples of which include Singapore, India, Botswana, Equatorial Guinea, Malta, Bulgaria and Brunei, have exhibited extraordinary innovation potential and have a long standing tradition of break-through inventions compared to the oil rich economies of the Gulf- which certainly have a wealth of creative people (De Boer and Turner, 2007). Here, a large number of family enterprises have excelled over the years in passing skills and knowledge to their children, reaffirming the work ethic and entrepreneurial spirit. Many Kuwaiti entrepreneurs are perceived not to be able to imagine life without the responsibilities of managing the family enterprise, because their working lives have revolved entirely around their enterprise, leaving little time to develop outside interests. However, cultural, economic and social transformations in the global markets will lead to a new demography affecting the way business is conducted. Indeed the promotion of all forms of innovation may be the single most lasting gift that one generation of Kuwaitis can grant on the next. (Palliam, Cader and Chiemeke, 2011).

4.7 Innovation in context

According to many, the process of innovation is the main engine of (continued) economic growth. As far back as 1943 Schumpeter (1943) emphasised that:

> "the fundamental impulse that sets and keeps the capitalist engine in motion comes from the new consumers' goods, the new methods of production or transportation, the new markets, the new forces of industrial organisation that capitalist enterprise creates."

However, such potential to create new products, processes, markets or organisations are path-dependent in the sense that there are certain nations and locations which seem to have acquired that capability over time, for innovation relies upon the accumulation and development of a wide variety of relevant knowledge (Dicken, 1998). The view that much needs to be in place for innovation to occur and that there is a significant role for the state to play is confirmed by George Marshall, whose ideas were responsible for the rebuilding of Europe after the Second World War. In the same publication he is cited as commenting on both the tangible and intangible aspects of the Industrial Revolution, suggesting that 'the secrets of Industry are in the air'. Marshall cited in Dickens (1998) recognised a number of characteristics that influence innovation:

- the institutional set-up;
- the relationship between the entrepreneurs and financiers;
- society's perception of new developments;
- the openness to science and technology;
- networks between scientific and academic communities and business circles;
- the operations of financial institutions;
- the growing liberal–individualist economic paradigm; and
- the role played by the state in accommodating and promoting.

However, the capability of organisations to initiate and sustain innovation is, to a great extent, determined by the wider local and national context within which they operate. This is essentially why 'innovation within' requires a favourable 'context outside'. That is, political, economic and social conditions play a major role in whether organisations or corporate actors will take the risk and innovate, as innovation is key to competitiveness, survival and sustained growth. To gain a better understanding of this, it is also necessary to 'look out of the window' at the business environment in which economic actors strive to gain an upper hand in the marketplace.

Much can be learned from examining recent history. The development of science and technology in the West created a wide gap between the so-called industrialised nations and their followers, 'late-industrialisers' (countries with no, or limited, indigenous technology development capacity). Some states, including Japan and some East Asian countries, have managed to close that gap with strategies which focus mainly on industrialisation. In these countries, economic growth was achieved in stages, through imitation, by diffusion of technology, development of new technology and efforts to develop their own capacities. So the cycle that began with imitation was later turned into a creative and broader basis upon which economic transformation could be achieved. This transformation required continual efforts by entrepreneurs and businesses and a collaborative framework promoted by the state. However, to reach maturity in today's economy (i.e. to be able to create high-value-added and knowledge-based products and services), would appear to be a gigantic task for the states and societies of latecomers. Consequently, the state must play a more significant role, perhaps by strategic intervention in infrastructure development and technological capacity formation, as well as investment in human capital formation.

4.8 The role of the state and national 'systems' of Innovation

To support our understanding of the process of innovation within the enterprise one must also grasp a basic understanding of the way the economy interrelates with global and regional economies on both local and national levels. Not only do national economies tend to be dominated by a form of economic organisation (e.g. the *Chaebol* in South Korea or *Keiretsu* in Japan), it is also the case that the relationship between state and business differs radically from one national space to the other. Kuwait, with its historic dependence on oil for example, is distinct from Japan – which does not possess oil or any substantive natural resources of note. Such interrelationships in society generate a business environment with a unique business value system, attitude and ethic. Historically, this difference created both advantages and disadvantages for business organisations across a range of activities, the most important of which may be perceived as the process of innovation (Porter, 1990).

The answer to the question of whether there is a role for the state in the process of innovation has been addressed in different contexts including those covered by Porter (1990) and Afuah (2003) and suggests state action may be necessary dependent upon a number of issues.

4.8.1 The 'public' nature of knowledge that underpins innovation

Knowledge is a type of public good, causing private expenditure on innovation to be dampened in those instances where the innovator cannot capture sufficient revenues to ensure an adequate return on their investment. In such scenarios, the government can play an indispensible role in the process of idea generation. Subsidisation and widespread distribution of knowledge may stimulate economic actors to work on new ideas, alongside state organisations, and to convert such ideas into marketable goods or services. For instance, by granting intellectual property rights to producers of knowledge and by establishing the necessary legal infrastructure to support those rights, the state may promote knowledge generation.

4.8.2 The uncertainty that often hinders the process of innovation

Macro-economic, technological or market uncertainties may all hinder innovation. When the environment causes companies to become risk-averse in investing funds in innovation projects, then the state may promote such activities through subsidisation, providing tax advantages and supporting firms in Research and Development projects. Creating a stable economic environment, where funds can be extended by the banking system to productive firms, is one of the first preconditions of strategy making. Thus, expectations of low inflation, low interest rates and stable growth will encourage firms to invest in entrepreneurial activity. Countries that exhibit political volatility manifest several forms of uncertainty and a reluctance to innovate. Auty (1994b) also demonstrates this in his classical study on Bolivian development. Good governance therefore becomes a precondition for progress of any kind in development. However, policymakers are usually frustrated by inadequate research on the causes, consequences and ultimate solution for poor governance particularly in developing economies. Uncertainty as a result of poor governance makes strategies for capital formation in developing nations more difficult.

4.8.3 The need for certain kinds of complementary investments

Provision of electricity, roads and water has historically assisted industrial development. More recently, the establishment of communication systems (e.g. communication superhighways), legal infrastructure and the formation of industrial districts (or clusters) have been areas where state action has established a favourable environment for enterprises.

4.8.4 The need for cooperation and governance

The state may lead in setting the vision, thus enhancing the possibilities for better communication and joint decision making. In the UK, for example, the government is providing funds (through education and promotion) to encourage households to switch from an analogue television signal to a digital television signal. Such action helps countries/society to embrace newer improved technologies more rapidly.

4.8.5 Politics

Lastly, in terms of politics, national states still play a key role in contributing to international and regional regulatory standards within the system of 'national states', and in creating consent and cohesion in the national business making arena. Such standards, whether environmental, safety or human rights, are increasingly becoming 'de rigueur' in industrial or business activities. The German government, for example, has an impressive record of being at the forefront of introducing legislation in automobile safety and environmental recycling, and this has contributed to Germany becoming a world leader in these two industries.

4.9 How national states can facilitate innovation

As both a financier of research and development and major purchaser, the state can have a significant impact on strategic direction within certain industries. In so doing it encourages entrepreneurial spirit. For instance, in 1995, the United States spent $71.4 billion on research and development in the defence, health, space, general science, energy, transportation, environment and agriculture sectors. There are also indirect ways of financing research and development, such as offering tax exemptions, subsidies, loan guarantees, and export credits. For example, Boeing not only paid no taxes between 1970 and 1984, but also received a tax refund amounting to $285 million (Afuah, 2003). As a major purchaser, the state can also reduce uncertainty and create favourable cash flows for firms by its willingness to pay higher (monopolistic) prices for early models.

Through education, information dissemination, governance and other societal actions, the state can impact upon the way the society perceives discoveries and adapts new technologies. At the same time, the state can also create cohesion in society and make strategic interventions to promote, for instance, the formation of a highly qualified workforce. Interdependency between state and society may create a favourable national culture which welcomes scientific development, and removes the potential for conflict between leading sectors and traditional sectors, economic interests and social forces and cultural traditions and new trends. Regulation of competition is another critical area where the state can intervene, ensuring that monopolies (that can result in under-innovation) do not emerge, and so protect society against possible corporate abuse.

Figure 4.2 The role of the state in innovation

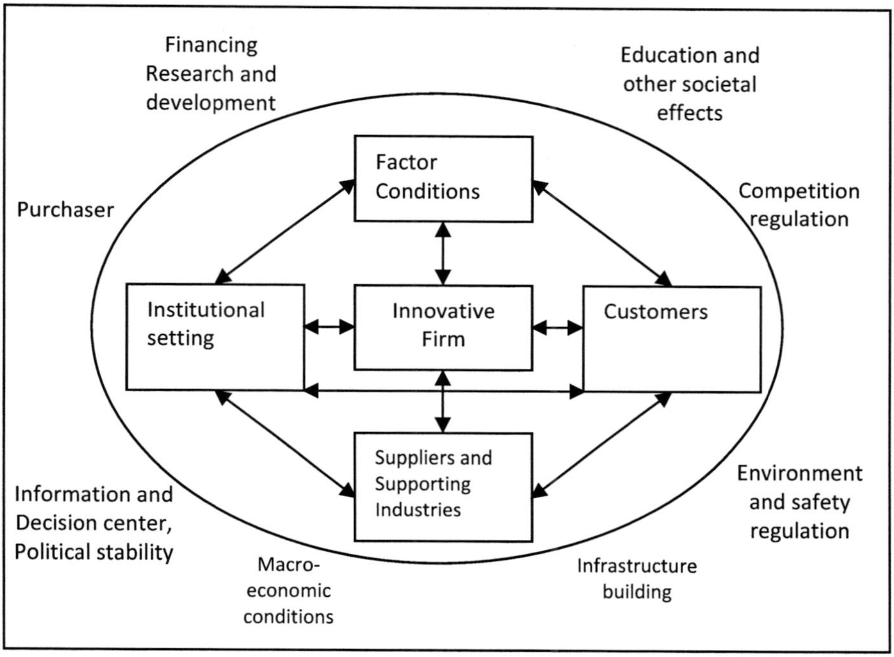

Source: Innovation Management and New Product Development (P. Trott, 2005)

Figure 4.2 highlights diagrammatically the possible roles that can be played by national states and the benefits that accrue to the various stakeholders. It takes Porter's industry attractiveness framework and examines the role the state can play in relation to innovation. It underlines a firm's relationship with buyers, factor inputs (land, labour, capital, raw materials); related and supporting industries (technology providers, input providers) and other institutions that help facilitate strategic orientation and innovative capabilities. These will determine to a great extent the firm's opportunities – notwithstanding the fact that its inner strengths (its strategy-making capabilities and structural features) will clearly affect this potential.

4.10 Kuwait's future depends on innovation

Globalisation has changed Kuwait as well as the world economic and social order. As a result new opportunities and new challenges have emerged. The opportunities are manifested in several growth metrics that have already been identified. Briefly, Kuwait is a wealthy economy and a geographically small state. The rise in global oil prices throughout 2010 has revived government consumption and economic growth as Kuwait experienced a 20% increase in government budget revenue. It is a relatively investor friendly and open economy and the crude oil reserves are estimated at 102 billion barrels – which constitutes approximately about 9% of world's reserve. Petroleum accounts for almost 50% of GDP, 95% of export revenues, and 95% of government income. The Kuwait Oil Company has an objective of increasing oil production to 4 million barrels per day by 2020. Possibly due to the positive fiscal situation, diversification of the Kuwait economy is minimal. The business climate is not highly considered and hostilities between the National Assembly and the executive branch have stalled several economic initiatives. Nonetheless, the government in May 2010 passed a privatization bill that allows the government to sell assets to private investors, and in January passed an economic development plan that pledges to spend up to $130 billion in five years to diversify the economy away from oil, attract more investment, and boost private sector participation in the economy. In any new economic order, Kuwait can only compete when a strategy of innovation has been implemented.

Al-Hassan, Delgado, and Omran, (2007) point out that while some Kuwaitis, concerned about issues ranging from climate change and the depletion of non-renewable resources to demographic change and emerging security needs (particularly after the invasion by Iraq), are calling for collective action to safeguard the Islamic way of life, others are advocating supporting future prosperity through social and economic development (with a key role envisaged for innovative ideas). From the protection of the environment through eco-innovation to the improvement of individual well-being through enhanced infrastructure provision, Al-Hassan, Delgado, and Omran (2007) strongly believe that innovation, in a broad sense, is one of the main responses to citizens' material concerns about their future. The key question that needs to be addressed is whether Kuwait has the extraordinary innovation potential seen in other Arab states which have a long-standing tradition of breakthrough inventions. For example science flourished under the early caliphs of Baghdad (Afsaruddin, 2002). By drawing on a variety of texts - Greek, Indian and

Persian - and translating them into Arabic, the early scholars accumulated a vast body of scientific knowledge in the world and built on it through their own discoveries. Astronomy was used to work out the direction of prayer. Mathematics was needed for dividing property according to the Islamic law of inheritance.

While increased competition may constitute the most efficient instrument to stimulate innovation, policy measures and innovation support mechanisms may also have a pivotal role to play. To facilitate innovation with sensitivity in Kuwait one needs to consider the establishment of problem solving, market directed and socially and environmentally responsible technology invention, commercial development and transfer systems.

The concept of 'developmental states' is used to explain the way in which some states achieved a major transformation of their economy and society. This was particularly the case with respect to some East Asian states, especially the so-called Tigers of Korea, Taiwan, Singapore and Hong Kong (Castells, 1992). Although such states were not immune to corruption, fraud and other forms of inefficiency, they brought about major changes in the economy, particularly in upgrading the potential of industry to move from imitation toward innovation and technology development, which is by no means an easy task. There are also 'predatory states', which capture most of the funds in the economy and reallocate them in the form of rents to a small group of the population, thus impeding the growth potential in the state (Evans, 1989).

4.11 The promotion and facilitation of technological innovation

The management of innovation has provided a large and diverse body of literature. It recognizes that while there is much complexity and uncertainty in managing innovation and new product development, much is known. There is considerable agreement on many of the factors that contribute to success and the activities and processes that need to be undertaken if innovation is to occur and a firm's performance is to improve. Table 4.2 presents some of the key studies that have influenced our understanding over the past fifty years.

This innovation management literature informs us that the capacity to absorb new knowledge, to transfer and diffuse knowledge and the ability to learn through interaction are crucial success factors in innovation as typified by Cohen and Levinthal (1989) Zahra and George (2002). New and commercially useful knowledge is not only the result of the conscious action of creative individuals. It is also the outcome of the interaction and learning processes among various actors in innovation systems. The need for connectivity, and the complexity of the interactions it entails, therefore emerges as a major factor influencing the management of innovation. Furthermore, Chesbrough (2003; 2003b; 2003c; 2006) argued recently that the process of innovation has shifted even further from one of closed systems internal to the firm to a new mode of open systems-involving a range of players distributed up and down the supply chain. This seems to be supported by the increasing application of network theory into more and more areas of business management (Parkhe et al., 2006).

Table 4.2 lists the major sources of secondary literature that were initially considered important to reflect upon innovation. It was not an intention to confine the study to these sources. However, the sources provide a historical account and detailed overview of innovation over the period 1957 – 2003.

The key studies in innovation management are presented in Table 4.2. It is the belief of the author that successful innovators encourage creative thinking and turn ideas into commercially viable products or services. Innovation can therefore be managed just like any other business discipline. Table 4.1 shows a historical discourse of innovation studies and this was initially adopted as a literature base for this chapter. The literature survey suggests that a firm's ability to successfully develop innovative new products is not only the result of public and private investments in tangibles and intangibles by individual elements in the economy, but that it is also strongly influenced by the character and intensity of the *interactions* between the elements of the system (see Figure 1.1, Chapter 1: Innovation systems and firm performance model).

Table 4.2 Key studies of innovation management

	STUDY	DATE	FOCUS
1	Carter and Williams	1957	Industry and technical progress
2	Project Hindsight- TRACES, (Isensen)	1968	Historical reviews of US government funded defence industry
3	Wealth from knowledge (Langrish et al.)	1972	Queen's Awards for technical innovation
4	Project Sappho (Rothwell, 1974)	1974	Success and failure factors in chemical industry
5	Von Hippel, E. (1976)	1976	The dominant role of users in the scientific instrument innovation process.
6	Stanford study (Madidique and Zirger)	1984	Success factors in US electronics industry
7	Minnesota Studies (Van de Ven)	1989	14 case studies of innovations
8	Rothwell	1992	25 yr review of studies
9	Sources of innovation (Wheelwright and Clark)	1992	Different levels of user involvement
10	MIT studies (Utterback)	1994	5 major industry-level cases
11	Project NEWPROD (Cooper)	1999	Longitudinal survey of success and failure in new products
12	Radical innovation (Leifer)	2000	Review of mature businesses
13	Christensen C M	2000	The innovator's dilemma
14	T.U.Delft study (Van der Panne et al.)	2003	A major literature review of success and failure factors.
15	Chesbrough	2003	Open innovation systems along the supply chain

Sources: Van der Panne et al. (2003) and Trott (2005)

This position is strongly advocated in the literature on "National Innovation Systems" by Freeman (1982), Lundvall (1992) and Nelson (1993). With this view, innovation and technological development in particular, depend increasingly on the ability to utilise new knowledge produced elsewhere and to combine this with knowledge already available in the economy and its actors. The capacity to absorb new knowledge, to transfer and diffuse knowledge, and the ability to learn through interaction are crucial success factors in innovation according to Cohen and Levinthal (1989) and Chesbrough (2003). New and commercially useful knowledge is not only the result of the conscious action of creative individuals; it is also the outcome of the interaction and learning processes among various actors in innovation systems, (i.e. producers, users, suppliers, public authorities, and scientific institutions), which David and Foray (1995) describe as the "knowledge distribution power" of the innovation system. The need for connectivity and the complexity of the interactions this entails therefore emerges as a major factor influencing the management of innovation. The development of network theory and network models of innovation have helped to illustrate further the prominence now given to internal and external networks within the innovation process (Dhanaraj and Parkhe, 2006; Parkhe et al., 2006). All these knowledge flows contribute to the wealth of knowledge held by the organisation (Woolgar et al., 1998; Rothwell, 1992; Major and Cordey-Hayes, 2002).

More recently, Chesbrough (2003) adopting a business strategy perspective, presents a persuasive argument that the process of innovation has shifted from one of closed systems (internal to the firm), to a new mode of open systems (involving a range of players distributed up and down the supply chain). Critics may argue that network models of innovation that emphasise external linkages have been around for many years (Taguchi and Nonaka, 1995; Trott, 1995; 2005). Significantly however, it is Chesbrough' emphasis on the new knowledge-based economy that informs the concept "open innovation". In particular it is the use of cheap and instant information flows which places even more emphasis on the linkages, and relationships of firms. It is from these linkages and the supply chain in particular, that firms have to ensure that they have the capability to fully capture and utilise ideas. Furthermore, the product innovation literature, in applying the open innovation paradigm, has recently been debating the strengths and limitations of so called 'user toolkits' which seem to ratchet up further this drive to externalise the firm's capabilities to capture innovation opportunities (von Hippel, 2001).

4.12 A review of models of innovation

The fundamental question for innovation research is to explain how firms develop innovative solutions in the form of products, processes and services. For many years this was seen as impossible to do, indeed it was viewed as a random phenomenon or "manna from heaven". In his early works, Joseph Schumpeter (1934) rejected this view and argued that innovation had three main aspects:

A high level of uncertainty;
The need to move quickly, to reap potential economic reward; and
Recognised the inherent resistance to new things.

This early view of innovation is sometimes referred to as Schumpeter Mark I. At this stage there is a clear emphasis and recognition of the need of the entrepreneur to fight numerous battles: the battle to get investors and secure their backing; the battle with competitors; and the battle to convince equivocal customers. Significantly, as models of innovation have developed over the twentieth century, the need to fight and battle seems to have been overshadowed by a growing list of factors deemed necessary for innovation to occur (Rothwell, 1992; Van der Panne, 2003). However, this misses the point about human involvement, and in particular the dynamic and crucial role of the entrepreneur. This study illustrates that firm innovation success today continues to be dependent on the battling activities of the entrepreneur, and these need to be central in any model of firm innovation.

Today, we recognise how different theoretical perspectives can shed light on different aspects of firm innovation, hence the growth in cross-disciplinary approaches to the study of innovation. (Evidence of this is provided by the research undertaken over the past thirty years at the Science Policy Research Unit, SPRU at the University of Sussex). In terms of developing understanding at the level of the firm, it is the research activities of business schools around the world that have added value. Traditional economics treated innovation as an activity that had inputs and outputs, and it was these that formed the basis of their studies. Hence, the focus on investment, allocation of resources, patents and profits – all easily quantifiable attributes. Indeed, until the 1960s, research focused on attempting to assess how much money needed to be spent on Research and Development to encourage innovation and what was its impact. In 1963, industrialised countries followed the US National Science Foundation and adopted the Frascati manual which offered helpful guidelines to facilitate measurements of science and technology (Godin, 2003).

Since then, largely because of its simplicity, this model has taken a firm grip on people's views on how innovation occurs. It is a linear model because there is a well-defined set of stages that innovations are assumed to pass through. Indeed, even the so-called ‘open innovation’ model of innovation essentially describes a linear process from research and development to marketing, with the innovation process being a funnel that begins with scientific research and progresses (linearly) via technological and product development to the market (Chesbrough, 2003). The problem with this view is that it generalizes a chain of causation that research only suggests for a minority of innovations, and it ignores the many feedback loops that occur between the various stages of the process (Kline and Rosenberg, 1986).

It was not until the 1980s that management schools around the world seriously began to challenge the sequential linear process. The recognition is that innovation occurs through the interaction of the science base (dominated by universities and industry), the technological base (dominated by Research and Development-intensive industry), and the needs of the market (dominated by innovative firms) was a significant step forward (see Figure 4.2) - and the explanation of the interaction of these activities forms the basis of contemporary models of innovation. However, there is of course much debate and contemporary disagreement about precisely what activities influence innovation (and, more importantly, the internal processes that affect a company's ability to innovate).

The interactive model emphasises that innovations occur as the result of the interaction of the market place, the science base and the organisation's capabilities(Figure 4.3). The use of information flows is used to explain how innovations emerge and how they can arise from a wide variety of points. Consequently, the overall innovation process can be thought of as a complex set of communication paths whereby knowledge is transferred. These paths include internal and external linkages. The generation of ideas is shown to be dependent upon inputs from three basic components (as outlined in Figure 4.3): organisation capabilities, the needs of the market place and the science and technology base (Bessant et al., 2000; Katz, 2002).

Figure 4.3 Conceptual chain of knowledge flows within innovation

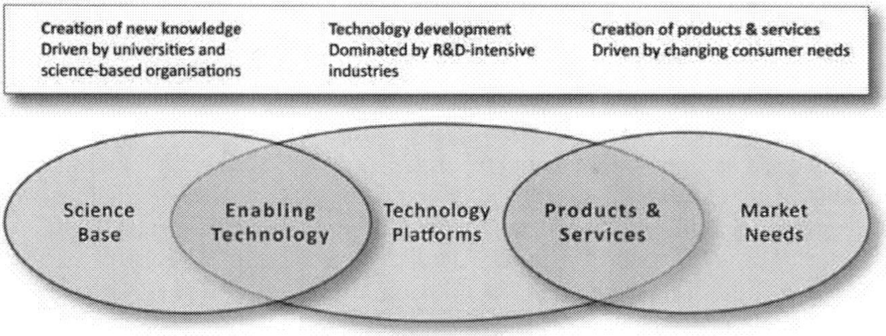

Source: Innovation Management and New Product Development (P. Trott, 2005)

According to Berkhout (2006) innovation is more than just a technical invention: economic, social and cultural aspects are often decisive. He goes on to argue that innovation is a symbiosis of these components. The preceding discussions also shows that innovation is not a singular event, but a series of activities that are linked in some way to the others. This may be described as a process according to Kelly and Kranzberg (1978), and involves:

i. A response to either a need or an opportunity that is context dependent;
ii. A creative effort which, if successful, results in the introduction of novelty; and
iii. The need for further changes.

The recognized complexity of the innovation process not only makes it difficult to manage, explain and study, but also presents problems for the development of a generalized model.

It is clear then that the innovation process is complex and full of uncertainties, which combine to make innovation a risky investment. However, the level of risk correlates to the level of improvement or change required. Scientists and engineers know they can improve the performance of virtually all existing products. The difficulty lies in the cost of achieving that improvement and also whether the market is willing to pay for that improvement. Furthermore, the process is dependent on information and knowledge flows, and it is this concept in particular which is a cornerstone of all models of innovation (Adams et al., 2006). It is also applicable to large and small firms alike (Berry and Taggart, 1994). This process is no different in Kuwait, as the next section shows.

4.13 Innovation as a driver of economic growth in Kuwait

An economic policy seeks to alter the structural conditions of a country's economy so as to create new opportunities for national economic development. Development is based upon two basic inseparable principles – economic growth and social development. Economic indicators show that the Kuwaiti economy recently witnessed a significant resurgence, because of the important and well-documented developments in the world's energy markets.

Kuwait posted huge, fiscal surpluses, reaching 5.2 billion dinars in the financial year 2006/2007, and 5.80 billion in the financial year 2008/2009. These developments have been reflected in the overall performance of the national economy, through a strengthening of economic activity rates in both the oil and non-oil sectors. As a result of this resurgence, average levels of personal income have improved from 4,902 dinars in 2002 to around 11,000 dinars in 2009, adding impetus to support the policies of economic prosperity initiated more than three decades ago.

These huge surpluses also allowed important initiatives to be taken by the State institutions, initiatives which are likely to have positive effects on growth for the future. These were epitomized by the efforts exerted by the Supreme Council for Planning and Development in laying down the foundations for indicative, developmental planning. A clear vision of growth for the State was drawn up, reflecting first and foremost the basic requirements outlined in His Highness the Amir's directives and aspirations, and specifying this vision's strategic goals thereby became the basis for the current growth plan. By virtue of these reforming initiatives, and in light of the recent favorable economic conditions, it was possible to lay firm foundations which will enable the Kuwaiti economy to diversify its product base and enhance its competitive capabilities. In starting to transform the State of Kuwait into a regional, commercial and financial centre, this will pave the way for a permanent launch-pad for the development of the economy and society as a whole. These ambitions exist, especially in light of the rapid development of neighboring states in the area, and in the context of untapped, social and economic development options, as the speed with which high rates of economic growth were achieved in previous years was not matched by similar growth in social and economic sectors.

14.3.1 The growth of Kuwaiti gross domestic product

The State of Kuwait benefited from the price rises in the world's oil markets during 2008 and this led to an increase in GDP. However, as Table 4.3 shows comparative growth rates in the oil sector were substantially higher than in the non-oil sectors over both periods (2002 – 2006 and 2006 – 2009). The data and reports produced by the Kuwait Central Statistics Office (2009), the Arab Planning Institute and the National Bank of Kuwait all suggest that the continuing dependence on the oil sector to drive economic activity represents a major challenge, and Kuwait should endeavour to gradually reduce this dependence by promoting growth strategies in the non-oil sectors.

Table 4.3 Sectoral growth rates in GDP (2002-2006) and (2006 –2009)*

Sector	**Nominal rate Growth %**	
Oil Sector:	**(2002 – 2006)**	**(2006 – 2009)**
Production of Crude and natural Gas	37.6	39.7
Manufacture of Refined products	28.3	27.3
Total Oil Sectors:	**36.7**	**42**
Non Oil Sectors:		
Financial services, Property, Work services	27.0	31
Society, Social and Personal services	8.2	7.3
Wholesale and Retail Trading, Restaurants/Hotels	7.5	9.1
Transport, Storage and Communication	24.4	23.3
Transferable Manufacture	14.4	10.2
Building-Construction	11.4	15.2
Electricity, Gas and Water	5.2	6.7
Agriculture, Livestock and Fishing	5.7	5.2
Total Non-Oil Sectors:	**15.9**	**17.3**
GDP at Purchaser's Price:	**25.5**	**27.8**

Source: based on data obtained from: Arab Planning Institute (API) http://www.arab-api.org, National Bank of Kuwait (http://www.nbk.com, Kuwait Central Statistics Office (2009) Statistical Review – Department of Publications and Dissemination.

Only the financial services, property and work services posted growth rates remotely close to those evinced in the oil sector, indicating that Kuwait has become ever more dependent upon the oil sector as a driver of growth and development. For innovation to be successful the dependence on the oil sector needs to be reduced and enhanced growth in the non-oil sector becomes necessary. Linked to this is the role of the private sector in fomenting such growth and by extension, innovation.

4.13.2 The private sector, GDP and innovation

For the purposes of growth, the increase in the level of participation of the private sector in economic activity is considered as one of the most important indicators of an economy's dynamism and development. The private sector, in its effort to maximise profit, makes use of creativity, inventiveness and innovation to meet market demands. When the role of the private sector is enlarged, it leads to the public sector relinquishing its role as the dominant force in economic activity, transforming its role to that of a regulator and a monitor (and a provider of a limited group of basic non-profit making services which the private sector fails to provide).

Although results indicate that the private sector's participation in GDP fell during 2002-2009 (from 37% to 30%, averaging 33% - as in Table 4.4), this is explained by the sharp growth in the oil sector over this period. If, however the oil sector is excluded, the importance of the private sector has been growing in recent years and it now accounts for almost two-thirds of non-oil GDP generation (Table 4.4). The rate of growth of the private sector during the last five years was not unexceptional (17.8 % annual rate of growth on average), and offers hope that private entrepreneurs have the capacity to drive growth and sponsor innovation in the Kuwaiti case.

With regard to the sectoral split within the activities of the private sector (Figure 4.4), 48% of activity is focused on three principal sectors: finance, insurance and real estate, and work services (which may be regarded as a start-point for transforming Kuwait into a regional financial centre). The transport and communications sectors represent only 10.6% of the total contribution to GDP of the private sector, an issue that needs to be reconsidered if these sectors are to make a more effective contribution in transforming Kuwait into a major commercial centre. The remaining private sector activities also remain small, transferable industries including the private oil industry, for example, represented only 9.3% of private sector output during the period, a result of the dominance of the public sector and the absence of sufficient incentives offered to the private sector in this important area of commercial activity.

Table 4.4 Public and private share of GDP, at current prices, for the period (2002-2009)

Sectors	Relative Importance in GDP for the period (%)	Rate of Annual Growth for the Period (%)
Public sector	67	25.4
Private sector	33	17.8
	100	23.5

Source: based on data obtained from: Arab Planning Institute (API) http://www.arab-api.org, National Bank of Kuwait (http://www.nbk.com, Kuwait Central Statistics Office (2009) Statistical Review – Department of Publications and Dissemination.

Figure 4.4 Sectorial distribution of the private sector output (2002-2006)

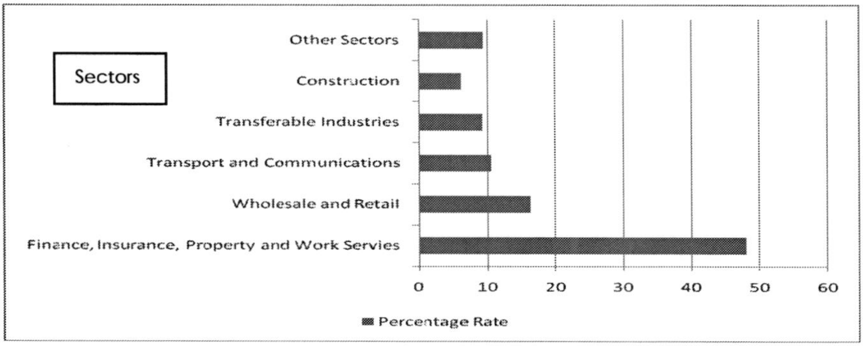

Source: based on data obtained from: Arab Planning Institute (API) http://www.arab-api.org

4.13.3 Scale of investment necessary for innovation

Recent years have witnessed a significant rise in total capital formation as a result of expenditure increases on investment projects - whether by government or the private sector. Overall, and as is evident from Table 4.5, expenditure on total capital formation registered an annual growth of 22% on average (2002-2008). The value of capital expenditure more than doubled in less than six years, to reach around 5,302 million dinars in 2008. These indicators confirm a state of recovery in investor confidence in the economy, and the Kuwait Central Statistics Office (2009) recorded increasing levels of demand for expatriate workers (non-domestic).

The private sector's share of total investments reached 45% on average during the period 2002-2006 (40% 2002 and 46% in 2006). On analyzing the components of private sector activity, the Kuwait Central Statistics Office (2009) found that 80% of total private investment was directed at only two sectors; the finance, insurance, real estate and work services sector (45% of investment), and the transportation and communications sector (35% of private sector investment).

Table 4.5 Total capital formation (2002-2008) at current prices

Period	**Value of Total Formation Capital (million dinars)**	**Percentage Annual Growth**
2002	1,985.1	--
2003	2,373.2	19.6
2004	3,185.5	34.2
2005	3,876.7	21.7
2006	4,366.4	12.6
2007	4,824.8	10.4
2008	5,301.3	10
Average for the Period 2002-2008		**16.34**

Source: based on data obtained from: Arab Planning Institute (API) http://www.arab-api.org, National Bank of Kuwait (http://www.nbk.com, Kuwait Central Statistics Office (2009) Statistical Review – Department of Publications and Dissemination.

Despite these positive developments, there are still some negative, structural phenomena. Amongst the most important is the continued dominance of final consumption in GDP. Table 4.5 and Figure 4.5 outline that the relative importance of capital formation remained relatively constant during 2002-2008, averaging around 16.86% (as opposed to around 55.75% for final consumption). These percentages are low when measured against international criteria, and could have a negative impact on the future growth of the local economy.

Table 4.6 Relative importance of capital formation in GDP (compared with other expenditure) at current prices (2002-2008)

Type of Expenditure/Relative Importance %	2002	2003	2004	2005	2006	2007	2008	Average for the Period
Final Consumption	74.9	65.7	57.3	47.9	43.2	47.6	53.7	55.75
Formation Total Capital	17.1	16.6	18.2	16.4	15.2	18.7	15.8	16.86
Exports of Goods and Services	44.6	52.1	56.9	64.0	66.5	60.9	65.7	58,67
Imports of Goods and Services	36.6	34.5	32.4	28.3	24.9	28.7	29.3	30.67
Expenditure on GDP	**100**	**100**	**100**	**100**	**100**	**100**	**100**	**100**

Source: based on data obtained from: Arab Planning Institute (API) http://www.arab-api.org, National Bank of Kuwait http://www.nbk.com, Kuwait Central Statistics Office (2009) Statistical Review – Department of Publications and Dissemination.

Figure 4.5 Expenditure on GDP

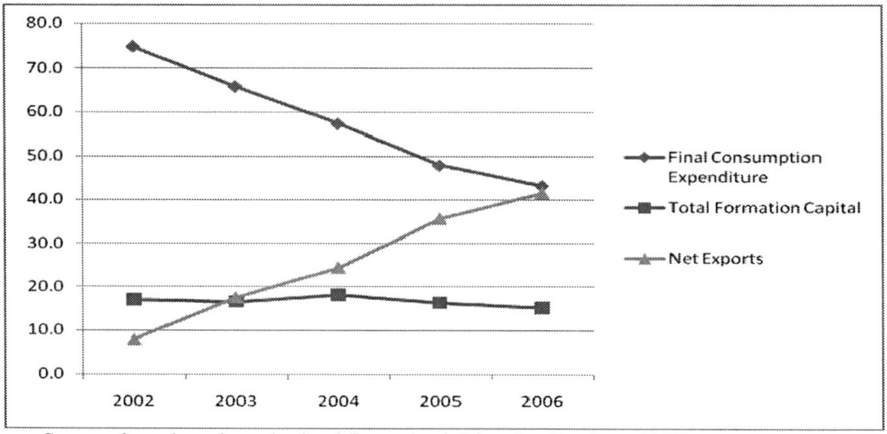

Source: based on data obtained from: Arab Planning Institute (API) http://www.arab-api.org, National Bank of Kuwait http://www.nbk.com, Kuwait Central Statistics Office (2009) Statistical Review – Department of Publications and Dissemination.

Data provided by the Kuwait Central Statistics Office (2009) suggest that a mismatch between consumption and capital formation is one of the phenomena which Kuwait aims to deal with within its current development strategy. This will be done through a series of initiatives aimed at raising the ability of the national economy to absorb a larger part of its savings (as a large proportion of this is currently invested outside the geographical boundaries of Kuwait by the Kuwait Investment Authority).

In this context, identifying metrics of innovation to better promote domestic growth and development becomes paramount. It is no good simply stating greater domestic investment/capital formation is necessary without assessing whether such investment can be effectively absorbed. Quality data and relevant and timely innovation metrics enhance public understanding, help policy-makers benchmark innovation performance and improve policymaking and business strategies, and also provide for the better utilization of domestic savings.

4.13.4 The growth of gross domestic product

The State of Kuwait benefited from the price rises in the World's oil markets and, at current prices, the average annual change in GDP reached 25.5% during the period from 2002 until 2006. An examination of this change in GDP makes it clear that the rate of growth varied, rising from 23% in 2003, 22.8% in 2004 and 54.7% in 2005 before falling back to around 21.5% in 2006.

The oil sectors (production of crude oil and natural gas and the manufacture of refined oil products) grew most in recent years. Having fallen in 2002, the rate of growth rose from 32% in 2003 to 50% in 2005 before dropping to 27% in 2006. As for the value added from the entirety of the non-oil sectors, the rates of growth varied during these years. The average rate of annual growth in the non-oil sectors (15.9%) represented only half the average rate of growth in the oil sectors (36.9%). This is attributable to the rise in oil prices during this period, and an increase in the quantities of crude oil produced (and the quantities of refined oil products exported).

Table 4.7 Development of the structure of GDP Sector (2002-2006)

Sector	Nominal Growth rate	Relative Importation rate to overall Product %
Oil Sector:		
Production of Crude and natural Gas	37.6	46
Manufacture of Refined products	28.3	4
Total Oil Sectors:	**36.7**	**50**
Non Oil Sectors:		
Financial services, Property, Work services	27.0	17
Society, Social and Personal services	8.2	15
Wholesale and Retail Trading, Restaurants/Hotels	7.5	5.5
Transport, Storage and Communication	24.4	5
Transferable Manufacture	14.4	3.6
Building-Construction	11.4	2.0
Electricity, Gas and Water	5.2	1.5
Agriculture, Livestock and Fishing	5.7	0.4
Total Non-Oil Sectors:	**15.9**	**50**
GDP at Purchaser's Price:	**25.5**	**100**

Source: based on data obtained from: Arab Planning Institute (API) http://www.arab-api.org

Figure 4.6 Oil/Non-Oil Domestic Product

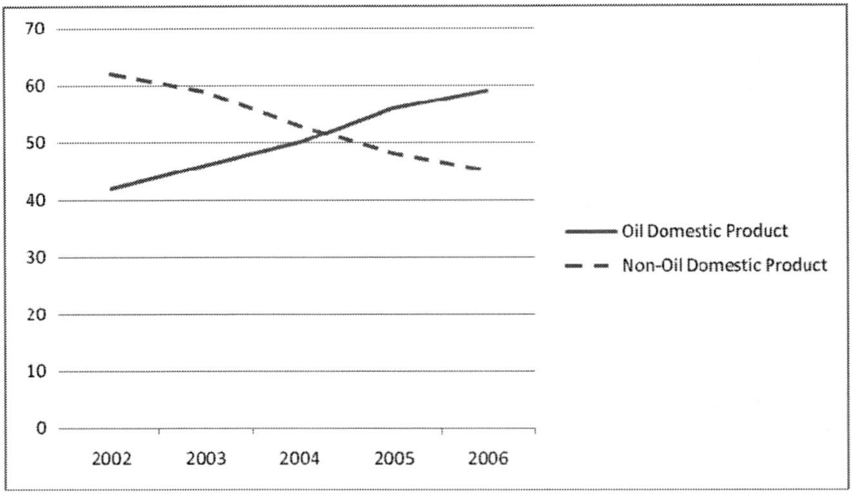

Source: based on data obtained from: Arab Planning Institute (API) http://www.arab-api.org, National Bank of Kuwait http://www.nbk.com, Kuwait Central Statistics Office (2009) Statistical Review – Department of Publications and Dissemination.

The continuing dependence on the oil sector to drive economic activity represents a big challenge which Kuwait is endeavoring to gradually overcome via its growth programmes for the forthcoming period; diversifying the economy's production base in a way that reduces this dependency in the future is considered to be Kuwait's primary, strategic aim. It is also considered to be one of the components for achieving a future vision for the State of Kuwait by transforming it into a regional, commercial and financial centre which is at the forefront of innovative activity.

4.13.5 The private sector, GDP and innovation

For the purposes of growth, the increase in the level of participation of the private sector in economic activity is considered as one of the most important indicators of an economy's dynamism and development. The private sector, in its striving to maximize profit, makes use of creativity, inventiveness and innovation to meet market requirements. When the role of the private sector is enlarged, it leads to the public sector relinquishing its role as the dominant force in economic activity, and transforming its role to that of a regulator and a monitor (and a provider of a limited group of basic non-profit making services which the private sector fails to provide).

Although results indicate that the private sector's participation in GDP fell during 2002-2006 (from 37% in 2002 to 31% in 2006). Nevertheless, the rate of growth of the private sector during the last five years was good (20% annual rate of growth on average).

Table 4.8 Public and Private share of GDP, at current prices for the period 2002-2006

Sectors	Rate of relative Importance in GDP for the period %	Rate of Annual Growth for the Period %
Public sector	67	28.4
Private sector	33	19.8
GDP at the purchaser's cost	100	25.5

Source: based on data obtained from: Arab Planning Institute (API) http://www.arab-api.org, National Bank of Kuwait http://www.nbk.com, Kuwait Central Statistics Office (2009) Statistical Review – Department of Publications and Dissemination.

The disparity between the recorded rate of growth of the private sector on the one hand and the noticeable drop of its relative share of GDP on the other can be observed from Table 4.8 and Figure 4.7.

Figure 4.7 Private Sector's Rate of Growth for 2002-2006

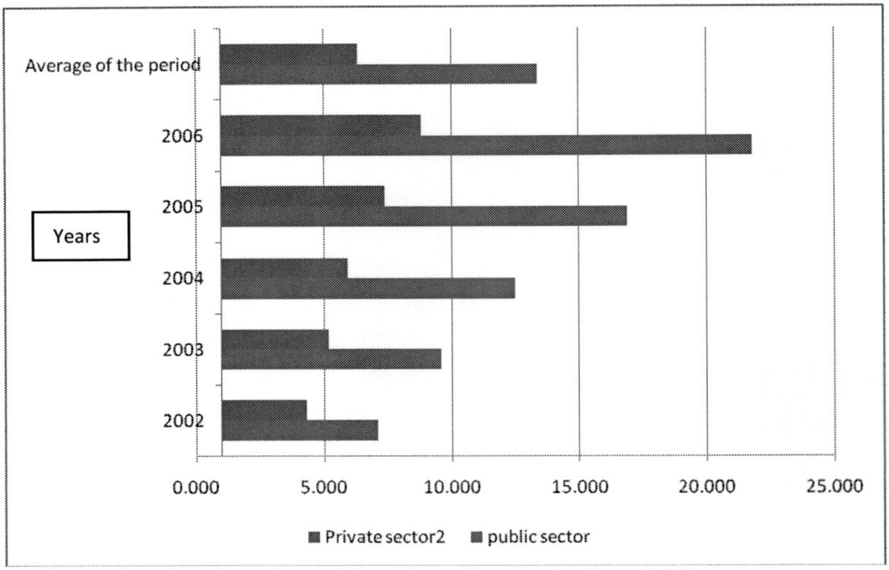

Source: based on data obtained from: Arab Planning Institute (API) http://www.arab-api.org, National Bank of Kuwait http://www.nbk.com, Kuwait Central Statistics Office (2009) Statistical Review – Department of Publications and Dissemination.

The reason is that the hike in World oil prices during this period was unmatched by a similar rate of growth in the output of the private sector. However, when one excludes oil, the private sector's contribution to non-oil GDP reached 59% on average during the same period (Table 4.9 and Figure 4.8). While the output value of the Kuwaiti private sector comes close to 4.3 billion dinars (which represents 56% of the non-oil GDP in 2002), it rose to close on 8.9 billion dinars by 2006 (62% of the non-oil GDP), that is to say more than double its value of 2002.

Table 4.9 Relative share of the private sector of Non-Oil GDP

Economic classification	2002	2003	2004	2005	2006	Average for the Period
Non-Oil Public sector (millions of dinars)	3,329.4	3,754.2	4,301.7	4,814.8	5,324.6	4,304.9
Private Sector (millions of dinars)	4,313	5,193	5,955	7,424	8,865	6,350
Non-oil GDP (millions of dinars)	7,642.4	8,947.2	10,257	12,239	14,190	10,654.9
Percentage share of the Private sector in non-oil GDP (%)	56	58	58	61	62	59

Source: based on data obtained from: Arab Planning Institute (API) http://www.arab-api.org, National Bank of Kuwait http://www.nbk.com, Kuwait Central Statistics Office (2009) Statistical Review – Department of Publications and Dissemination.

Figure 4.8 Private Sector Share of Non-Oil GDP for the period 2002-2006

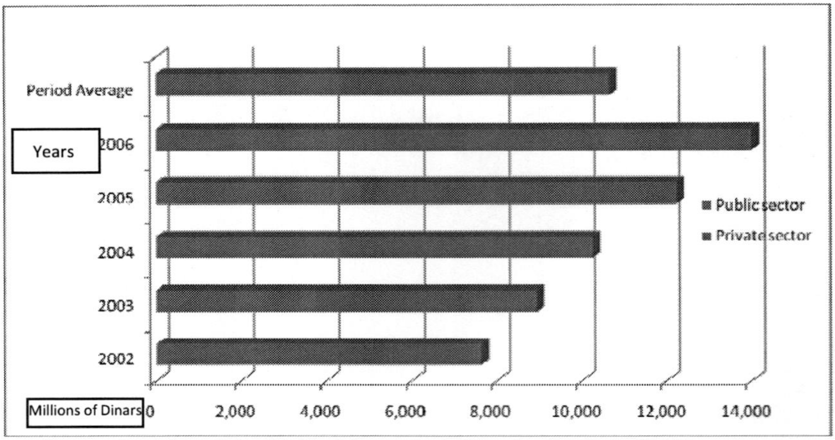

Source: based on data obtained from: Arab Planning Institute (API) http://www.arab-api.org, National Bank of Kuwait http://www.nbk.com, Kuwait Central Statistics Office (2009) Statistical Review – Department of Publications and Dissemination.

With regard to the sector split within the activities of the private sector, 48% of activity is focused on three principal sectors: finance, insurance and real estate, and work services (which may be regarded as a start-point for transforming Kuwait into a regional financial centre). The transport and communications sectors represent only 10.6% of the total contribution to GDP of the private sector, an issue that needs to be addressed quickly if these sectors are to make a more effective contribution to providing the basic components necessary for transforming Kuwait into a major commercial centre. The remaining private sector activities also remain small - Transferable industries - including the private oil industry -for example, represented only 9.3% of private sector output during the period 2002-2006 (Figure 4.9), a result of the total dominance of the public sector and the absence of sufficient incentives offered to the private sector in this important area of commercial activity.

Figure 4.9 Sector Distribution of the Private Sector Output

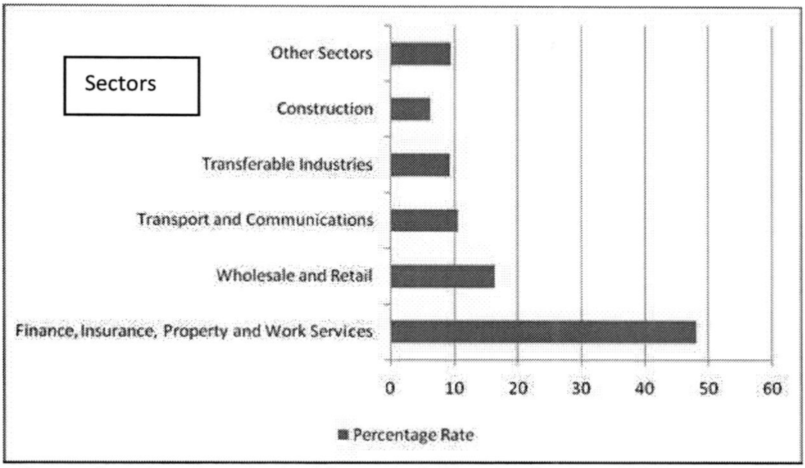

Source: based on data obtained from: Arab Planning Institute (API) http://www.arab-api.org, National Bank of Kuwait http://www.nbk.com, Kuwait Central Statistics Office (2009) Statistical Review – Department of Publications and Dissemination.

4.13.6 Inflation in Kuwait

The inflation rate is Kuwait is estimated to be around 4.5%. This has increased over the past 5 years and also from 2002 till 2007. Figure 4.10 shows the changes in inflation measured in terms of consumer prices and wholesale prices.

Figure 4.10 Development of inflation rate 2002-2007

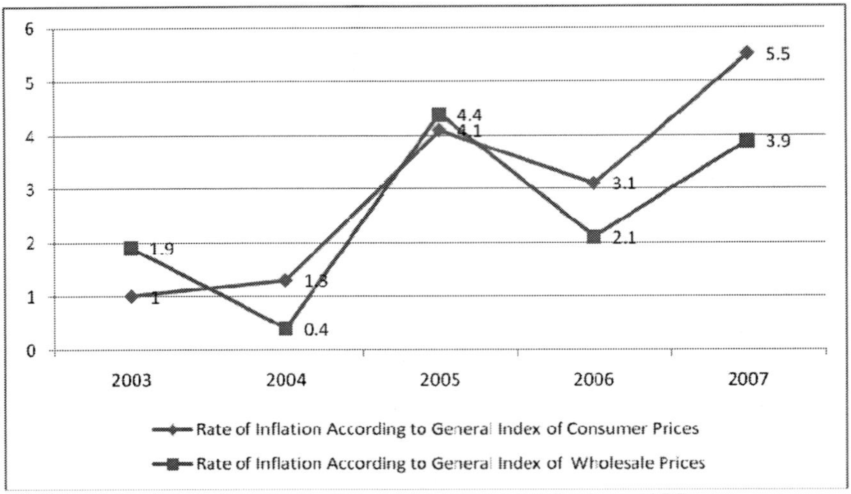

Source: based on data obtained from: Arab Planning Institute (API) http://www.arab-api.org, National Bank of Kuwait http://www.nbk.com, Kuwait Central Statistics Office (2009) Statistical Review – Department of Publications and Dissemination.

4.13.7 Scale of investment necessary for innovation

Recent years have witnessed a significant rise in the mode of expenditure on total formation capital as a result of expenditure increases on investment projects - whether by government or the private sector. The private sector's share of total investments reached 45% on average during the period 2002-2006 (40% 2002 and 46% in 2006). On analyzing the components of private sector activity, it is clear that 80% of total private investment was directed at only two sectors; the finance, insurance, real estate and work services sector (45% of investment), and the transportation and communications sector (35% of private sector investment).

Overall, and as is evident from Table 4.10, expenditure on total State capital formation registered an annual growth of 22% on average (2002-2006). The value of capital expenditure doubled in less than four years, to reach around 4,366 million dinars in 2006. These indicators confirm a state of recovery in investors' confidence in the economy, and we can now observe that the recorded levels of demand for expatriate workers (non-domestic) increased during the past three years.

Table 4.10 Development of total formation capital (2002-2006) at current prices

Period	Value of Total Formation Capital (million dinars)	Percentage Annual Growth
2002	1,985.1	--
2003	2,373.2	19.6
2004	3,185.5	34.2
2005	3,876.7	21.7
2006	4,366.4	12.6
Average for the Period 2002-2006	**3,157.4**	**22.0**

Source: based on data obtained from: Arab Planning Institute (API) http://www.arab-api.org, National Bank of Kuwait http://www.nbk.com, Kuwait Central Statistics Office (2009) Statistical Review – Department of Publications and Dissemination.

Despite these positive developments, there are still some negative, structural phenomena. Amongst the most important is the continued dominance of final consumption in GDP. Table 4.11 and Figure 4.11 outline that the relative importance of capital formation did not rise during 2002-2006, recording an average of around 16.5% (as opposed to around 55% for final consumption). These percentages are low when measured against international criteria, and could have a negative impact for the future growth of the local economy.

Table 4.11 Development of the relative importance of formation capital in GDP at current price for 2002-2006

Type of Expenditure/Relative Importance %	2002	2003	2004	2005	2006	Average for the Period
Final Consumption	74.9	65.7	57.3	47.9	43.2	54.1
Formation Total Capital	17.1	16.6	18.2	16.4	15.2	16.5
Exports of Goods and Services	44.6	52.1	56.9	64.0	66.5	59.3
Imports of Goods and Services	36.6	34.5	32.4	28.3	24.9	29.9
Expenditure on GDP	**100**	**100**	**100**	**100**	**100**	**100**

Source: based on data obtained from: Arab Planning Institute (API) http://www.arab-api.org, National Bank of Kuwait http://www.nbk.com, Kuwait Central Statistics Office (2009) Statistical Review – Department of Publications and Dissemination.

Figure 4.11 Expenditure on GDP

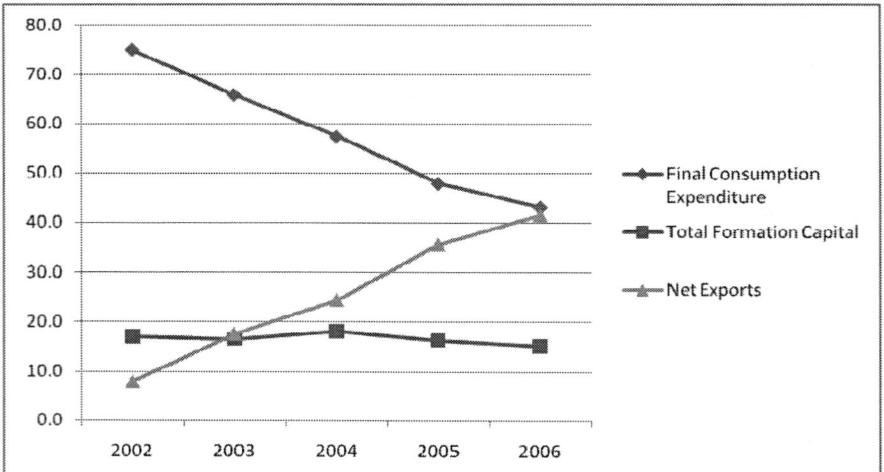

Source: based on data obtained from: Arab Planning Institute (API) http://www.arab-api.org, National Bank of Kuwait http://www.nbk.com, Kuwait Central Statistics Office (2009) Statistical Review – Department of Publications and Dissemination.

This mismatch in the relative relationship between consumption and capital formation is one of the phenomena which Kuwait aims to deal with within its development strategy. This will be done through a series of initiatives aimed at raising the ability of the national economy to absorb the largest part of savings capacity a large proportion of which is currently invested outside the geographical boundaries of Kuwait. Recorded results have shown that the proportion of total capital formation to net savings did not exceed 50% on average during 2002-2006.

4.13.8 Foreign trade

As is clear from Table 4.12 the rate of annual growth during this period was in the region of 38.8% (on average). On this basis, the value of exports at current prices doubled about four times between 2002 and 2006, and the proportion of exports to GDP reached approximately 59.3% (on average) during this period. In contrast imports at current prices during the same period grew at a rate of 13.9% on average (whilst the proportion of imports to GDP was around 30% on average).

Table 4.12 Development of Foreign trade at current prices (2002-2006)

Data	2002 Value	2003 Value (Growth %)	2004 Value (Growth %)	2005 Value (Growth %)	2006 Value (Growth %)	Average Value (Growth%)
Exports of Goods and Services	5.171 ---------	7.432 43.72%	9,970 34.15%	15,094 51.39%	19,038 26.13%	11,341 38.85%
Imports of Goods and Services	4,243 -------	4,917 15.88%	5,672 15.35%	6,670 17.60%	7,123 6.79%	5,725 13.91%

Source: based on data obtained from: Arab Planning Institute (API) http://www.arab-api.org, National Bank of Kuwait http://www.nbk.com, Kuwait Central Statistics Office (2009) Statistical Review – Department of Publications and Dissemination.

As a result of this clear disparity in the speed of growth of exports on the one hand and imports on the other, the Kuwaiti economy was able to bolster its excellent position vis-à-vis its foreign exchange balances. The trade balance rose from 928 million dinars in 2002 to 11,915 million dinars in 2006 (5,616 million dinars on average), with crude oil representing more than 90% of total Kuwaiti exports. At the same time the contribution of other exports and re-export activities remained very modest, undermining Kuwait's desire to transform itself into a regional commercial centre. An analysis of the structure of imports confirms the modest nature of their role in supporting the Kuwaiti economy, since basically they are made up of consumer goods and various service imports. Similarly, their increase in the proportion of GDP directs revenues more towards consumer spending than towards savings and investment and generally weakens the basic productivity structure in the local economy. The first, inclusive strategic goal is to diversify the economy's productivity base, establish a competitive economy and create tangible changes in the structure of GDP.

4.14 Innovation capacity in the Gulf and Kuwait

Wilson (2010) questions whether it is possible to build sustainable innovation in oil rich countries of the Gulf that are heavily dependent on wealth from a single resource. In this regard he ascertained that each of the six Gulf States is pursuing some economic diversification strategy to move their respective economies away from hydrocarbon dependence. Wilson (2010) and Palliam (2010) are consistent in their views over diversification and suggest that any move by Gulf countries towards diversification needs responses to the questions:

Firstly, how economically viable is the diversification measured in terms of global competitiveness?
Secondly, what are the best practices and to what degree the best practices are embodied in these new diversified sectors;
Thirdly, what is the role of innovation in this regard; and
Fourthly, would Gulf economies be able to sustain new diversified sectors particularly when wealth from one leading sector is used to leverage the new diversified sector.

Innovation and whether it is sustainable impacts directly upon the international competitiveness of a country. Palliam (2010) contends that competitiveness is the key to productivity growth and goes on to define competitiveness as the set of institutions, policies, and factors that determine the level of productivity of a country. Competitiveness is simultaneously driven by a combination of macroeconomic policies, a sound institutional framework, modern infrastructure and an efficient financial system to ensure an enabling business and investment climate. From this point of view Gulf countries seem to be more than adequate. However, one impediment to global competitiveness particularly of the new diversified sector is – to what extent is the economy open and accessible. Gulf countries are not homogeneous economies. However, one issue that is common is the lack of international competitiveness that emanate from several inadequacies (Wilson, 2010). In this regard Palliam (2010) critically attributes the lack of international competitiveness to inadequate institutions in the Gulf to address unnecessary bureaucracies; dysfunctional regulatory authorities; high prevalence of corrupt officials; and the low levels of innovation capacity and innovation systems as a result of a reluctance to build capacity.

To achieve best practices in any sector, particularly in a knowledge-based economy, investment in research and development is paramount. Wilson (2010) identifies the role of universities in this regard. Palliam (2003) argues that no economy can exist without a university to support and no university can exist without a good economy to sustain it. University and economy collaboration in research and development is a result of a long history of good governance, and coherent science, technology and innovation policies. The role of educational institutions in empowering individuals (human capital) with intellectual capital is also critical. Wilson (2010) goes on to conclude that Gulf States lag dramatically behind OECD countries in terms of research and development. Within this context Wilson (2010:5) contends that "there is very little scope currently for Gulf States to become more internationally competitive with respect to innovation and knowledge fundamentals".

To build a knowledge-based economy and society requires the development of innovation through: knowledge production; knowledge application and knowledge diffusion. Gulf States willingness and earnestness to diversify its narrow economic base are indeed the first signs of innovation and of course there are many initiatives to broaden the education base of Gulf States. Wilson (2010) cites the United Kingdom's Department of Trade and Industry which defines a knowledge-driven economy as one in which the generation and exploitation of knowledge play the predominant part in the creation of wealth. This in essence suggests that economic success of any country depends upon knowledge creation and learning. The OECD makes an identical assertion – that the term "knowledge-based economy" results from a fuller recognition of the role of knowledge and technology in economic growth. Intellectual capital in any form (natural or artificial intelligence) has always been central to economic development. For any successful economy to sustain innovation, it must be able to produce, distribute and use knowledge more than ever before. The ability to improve the economy's intellectual capital is indeed a source of competitive advantage.

The next related issue is how a knowledge economy measured – Knowledge Economy Index. For this purpose, Wilson (2010:6) identifies the use of a World Bank Institute Index that comprises of four pillars, namely:

Pillar One - Economic and Institutional Regime where the country's economic and institutional regime must provide incentives for the efficient use of existing and new knowledge and the flourishing of entrepreneurship.

Pillar Two - Education and Skills where citizens need education and skills that enable them to create and share, and to use it well.

Pillar Three - ICT Infrastructure where a dynamic information infrastructure is needed to facilitate the effective communication, dissemination, and processing of information.

Pillar Four – An Innovation system where the country's innovation system - firms, research centers, universities, think tanks, consultants, and other organizations - must be capable of tapping the growing stock of global knowledge, assimilating and adapting it to Iocal needs and creating new technology.

Table 4.13 presents the rankings of the Gulf States in terms of 146 countries that constituted the sample for the World Bank Institute Index. The Knowledge Economy Index for the Gulf States is provided in the accompanying table. The ranking of Kuwait in relation to the six Gulf States is also provided.

Table 4.13 The rank of Gulf Countries among 146 countries with regard to knowledge economy index (KEI)

Country	**KEI Overall**	**KI Diffusion**	**Pillar 1 Economic**	**Pillar 2 Education**	**Pillar 3 ICT**	**Pillar 4 Innovation**
Bahrain	49	56	48	60	40	80
Kuwait	52	59	51	76	46	70
Oman	66	79	40	86	76	71
Qatar	44	45	42	67	27	48
Saudi Arabia	68	73	58	80	52	86
UAE	45	44	47	79	21	46
Kuwait Ranking	**3rd out Six**	**4th out Six**	**5th out of Six**	**2nd out Six**	**4th out Six**	**3rd out of Six**

Source: Adapted from Wilson (2010) Is it possible to build sustainable innovation capacity in oil rich Gulf Countries? XXII ISPIM Conference, June 12- 15, 2010 – Hamburg Germany.

Overall Kuwait's ranking 52 out of 146 country augers well for the Gulf State. However, the factors that impede the creation and diffusion of a knowledge environment need to be taken more seriously. In terms of education Kuwait ranks 2nd out of six Gulf States. This is not surprising particularly when one considers that Kuwait has a large number of foreign educational institutions offering a variety of programs. Unfortunately, most of the private educational institutions are not in the public domain. Kuwait ranking 5th out of six, needs to address its economic and institutional infrastructure where Kuwait must create incentives for the efficient use of existing and new knowledge to ensure that entrepreneurship flourishes.

Generally Gulf countries are way behind in terms of educational and innovation fundamentals compared to competitiveness of other countries. It is important that countries in the Gulf bridge the substantial gaps. Due consideration must be given to whether Gulf countries would be able to sustain their extravagance with their current wealth status and at the same time ignoring competitive innovation systems and knowledge economy fundamentals. Economic diversification encompassing sustainable innovation systems and diversification enabling an environment of national innovation ecosystems are recommended by Wilson (2010). This recommendation is consistent with what Palliam (2011) refers to as issues in Gulf Security.

4.15 Results of the Public Finances

Recently Kuwait has seen significant growth in government revenues as a result of the growth in oil income. Government revenue growth exceeded growth in government spending, and the result was a large, financial surplus which came to the fore with effect 2002/2003. The value of this surplus reached around 670 million dinars in that year. It is evident from the following table that since then the surplus has continued to increase, reaching around 5,493 million dinars in 2005/2006 before falling back to around 3,650 million dinars in 2006/2007. Based on this data, the annual average surplus was in the region of 2,457 million dinars during the period 2002/2003 – 2006/2007, with oil revenues representing the equivalent of 91.2% of the total value of public revenues. As far as expenditure is concerned (and as is evident from the data in Table 4.14), current expenditures, as compared to capital expenditures dominate. Current expenditures reached around three times that of capital expenditures on average during the period (basically due to a rise in the proportion of public sector wages, salaries and internal transfers as the government employs more than 85% of the country's indigenous workforce).

Although the current financial situation is largely satisfactory (as these huge recorded oil revenues over the past five years have largely contributed to supporting the activities of the State), the outgoings are regarded as a drain on oil wealth. The wealth is being transformed to direct government spending instead of being invested in income bearing assets (which could finance government expenditures from their returns whilst preserving the assets). Equally, the domination of oil revenues – which is considered revenue ready to disburse – has weakened the basic role of other revenue sources. For example, direct and indirect taxes have made no real contribution (>2.2%) towards financing the budget of the State at any rate during the last five years.

Table 4.14 Development of the values and framework of revenues and public expenses (2002-2003)

Data	2002/2003	2003/2004	2004/2005	2005/2006	2006/2007	Average
Total Public Revenues						
Value	6,219	6,937	8,962	13,728	15,509.23	10,271
Rate of Growth	16.5	11.6	29.19	53.17	12.97	
Total Expenses						
Value	4,927	5,523	6,315	6,862	10,306.36	6,787
Rate of Growth	3.8	12.1	14.35	8.658	50.2	
Surplus	670	720.7	1,751	5,493	3,651.95	2,457
Relative distribution of the Public revenues Framework						
Oil Revenue	88%	89%	91%	94%	94%	91%
Non-Oil revenue	12%	11%	9%	6%	7%	9%
Total Public Revenues	100%	100%	100%	100%	100%	100%
Development of the Values and Relative Distribution of the Public Expenses Framework						
Salaries						
Value	1,542	1,637	1,754	1,931	2,225.89	1,818
%	31.2	29.6	27.8	28.1	21.6	27.7
Commodity Necessities and Services						
Value	11.8	668.2	870.3	1058	1,372.49	910.2
%	5.82	12.1	13.8	15.5	13.4	13.3
Transport, Equipment and supplies						
Value	23.7	40.6	44.5	58.64	77.08	48.9
%	0.5	0.7	0.7	0.8	0.7	0.7

Construction and Maintenance Projects						
Value	461	569.5	678.3	750.5	989.45	689.8
%	9.4	10.3	10.7	11	9.6	10.2
Various Expenses and Transformed Payments						
Value	2,319	2,608	2,968	3,064	5,641.45	3,320
%	47.1	47.3	47	44.6	54.7	48.1
Total Expenses Value	4,927	5,523	6,315	6,862	10,306.36	6,787
%	100	100	100	100	100	100

Source: based on data obtained from: Arab Planning Institute (API) http://www.arab-api.org, National Bank of Kuwait http://www.nbk.com, Kuwait Central Statistics Office (2009) Statistical Review – Department of Publications and Dissemination.

4.16 Innovation in the Commercial and Financial Service Sectors

Developing these two sectors represents the lynchpin if the State is to achieve its vision of transforming the country into a regional, commercial and financial centre. **The country's financial sector** has witnessed quantitative and qualitative developments during the past five years. Its rates of growth have risen and its assets have clearly increased (Table 4.15). This sector basically consists of local banks, Islamic, traditional investment companies, investment funds, insurance companies and the Kuwait stock exchange. As far as banks are concerned and as is evident in the following table, the value of their assets rose significantly in recent years from about 17 billion dinars in 2002 to around 35 billion dinars at the end of 2007 (an annual percentage growth of 22% on average).

Table 4.15 Value of local Banks and investment companies Assets 2002-2007

Data	2002	2003	2004	2005	2006	2007
Value of Bank Assets	17,063	18,813	19,144	21,611	26,990	35,552
(Rate of Growth)	-------	10.3%	1,8%	12.9%	24.9%	31.7%
Value of Traditional Investment Companies' Assets	3,465	4,236	4,868	6,471	7,569	9,386
(Rate of Growth)	-------	22.3%	14.9%	32.9%	17.0%	24.0%
Value of Islamic Investment Companies' Assets	880	1,291	1,488	2,482	3,568	6,634
(Rate of Growth)	--------	46.7%	15.3%	66.8%	43.8%	85.9%

Source: based on data obtained from: Arab Planning Institute (API) http://www.arab-api.org, National Bank of Kuwait http://www.nbk.com, Kuwait Central Statistics Office (2009) Statistical Review – Department of Publications and Dissemination.

Assets of investment companies registered with the Central Bank of Kuwait also enjoyed a huge rise for the period 2002-2007, from around 3.5 billion dinars to reach approximately 16 billion dinars (51% of which were traditional investments and 41% Islamic investments), an annual rate of growth of 54% on average. Assets of traditional investment companies also doubled to reach 9.4 billion in 2007 whilst assets of Islamic investment companies rose six times to a figure estimated at 6.6 billion dinars (Figure 4.12).

Figure 4.12 Assets of Kuwaiti investment companies 2002-2007

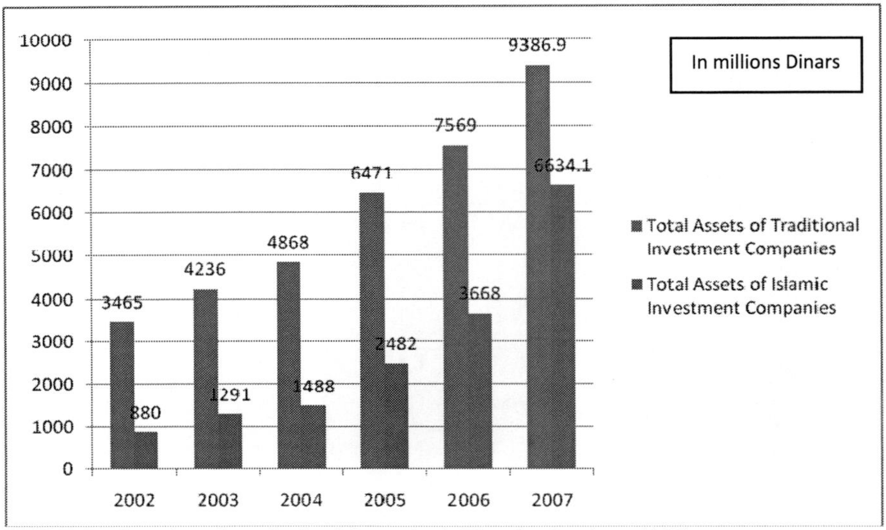

Source: based on data obtained from: Arab Planning Institute (API) http://www.arab-api.org, National Bank of Kuwait http://www.nbk.com, Kuwait Central Statistics Office (2009) Statistical Review – Department of Publications and Dissemination.

To complement the strong resurgence of the Kuwaiti financial sector, investment funds' assets witnessed a strong annual growth of 36% approximately during 2002-2006 (as reflected in Table 4.16). This resulted in a rise of asset values from around a billion dinars in 2002 to around 3 billion in the year 2006, with the number of investment funds increasing from 33 to 79. As for the insurance sector, despite its relatively limited capacity (compared with other countries), it also witnessed a significant growth in recent years. This can be explained as the Ministry of Commerce and Industry granted licences for the pursuance of insurance activity to several companies in 2004. During the period 2002-2006 their assets rose to 27% on average to reach almost 679 million dinars in 2006.

Table 4.16 Value of investment fund and insurance companies assets registered with the Central Bank during the period 2002-2006

Item	2002	2003	2004	2005	2006
Total Investment Fund Assets	1,016	1,686	2,342	2,868	2,843
(Rate of Growth)	------	66%	39%	22%	-1%
Total Insurance Company Assets	326.4	385.1	411.9	469.4	678.8
(Rate of Growth)	------	18%	7%	14%	45%

Source: based on data obtained from: Arab Planning Institute (API) http://www.arab-api.org, National Bank of Kuwait http://www.nbk.com, Kuwait Central Statistics Office (2009) Statistical Review – Department of Publications and Dissemination.

As for the Kuwait Stock Exchange, it has also witnessed clear growth in the numbers of registered companies and in its market value in recent years. The market, one of the biggest and most active in the Gulf and Arab region, had a market value reaching about 16.7 billion dinars in 2006 (6.7 billion in 2002). As evidence of the outstanding performance of the financial sector, Kuwait recently secured a risk grading of AAA, the highest valuation level given by the leading agencies of international credit classification (such as Standard andPoor's, Moodys, Capital Intelligence, the Fitsh Group and Investment Information). Despite these positive developments, growth in the financial sector and the realization of a positive rate of development in the next five years is conditional upon the sector's ability to confront the challenges which might impede future success. Kuwait has responded to this with "The Kuwait Report 2020: Transforming Kuwait into a Commercial and Financial Centre". This included 23 important initiatives for transforming the country into a specialist, regional financial centre. In this context the most important priority lies in dealing with the administrative and organizational obstacles which confront the sector, and in developing a legislative environment (including updating the legal framework for business operations, and implementing the principles of "good, organizational, management practice"°. It also includes the establishment of the "independent body for the financial markets" which will organize and oversee the market in a way that enables Kuwait to achieve the international standards laid down by the International Commission for Financial Markets, and thereby raise the level of competition in the banking sector.

The commercial sector has continued to play a limited role in the economic activity of the State. According to the available data, the contribution of the wholesale and retail commercial sector to non-oil GDP was around 8.5% on average. Data from the Kuwait Chamber of Commerce and Industry shows that this sector recorded significant growth in value during the last two years, and resulted in a rise of expenditure which exceeded the billion dinar barrier for 2005 and 2006 (1.01 and 1.09 billion respectively). Furthermore, the wholesale and retail sector (plus restaurant and hotel operations) absorbed the largest part of the workforce in the Kuwaiti economy - the number of workers in this sector reaching around 481 thousand at the end of 2006 (which is equivalent to 45.4% of the total workforce).

These developments are a reflection of the recent rise in economic growth rates and the resultant rise of the individual's share of GDP (due to an increase in demand for consumer goods). They are also a reflection of the availability of an independent, civil judiciary to resolve commercial disputes. In this context, Kuwait has striven in recent years for more trade openness, and for a greater strengthening of commercial relations with a number of states (both regionally and globally). In 2005 custom tariffs on imports originating from the member states of the Greater Arab Free Trade Organization were revoked and in the same year a number of bi-lateral agreements were signed with a number of states to avoid tax duplication (South Africa, South Korea, Singapore, Malaysia and Hong Kong). During this period, Kuwait also signed a general framework agreement for trade and investment with the USA as a first step towards developing free trade agreement between the two countries.

In their entirety, these measures are initial steps which will contribute towards strengthening Kuwait's strategic geographical status, allowing it to perform a pivotal role in commercial goods trade. However, facing up to the challenges which confront the commercial sector are imperative in order to achieve Kuwait's vision as a commercial centre. The most important of these challenges is in the area of logistics. The initiatives aim to achieve greater development of the commercial infrastructure; including warehousing and Customs depots, an increase in frozen storage capacity and facilities to preserve and handle perishable goods. The plan is also expected to monitor the commercial sector so as to eliminate commercial cheating and manipulation and to guarantee the protection of the consumer. The organizational and legal environment will also require form, especially the laws restricting commercial activity, the obstacles which stand in the way of issuing licenses to companies, and the full gamut of customs procedures.

4.17 Recorded Results of Kuwait's Competitive Ability

Competitiveness is considered as one of the components for analyzing the health of an economy's performance. By using a group of indicators we can measure the degree of economic policy effectiveness and the government's capability of enabling national, economic entities or regional partnerships to resist foreign, competitive pressure within the Kuwaiti market. These indicators can also measure the ability of these economic entities to penetrate neighbouring and international markets. The latest global competition reports indicate that Kuwait's position on the business competition index - which lists the states of the World-wide report for competition - is 44 out of a total of 126 states for the year 2006-2007. Similarly Kuwait occupied the eighth position in the report of 10 states (amongst those states sampled by the National Kuwaiti Competitiveness Commission; Table 4.17) after Qatar and the Emirates. The business competition index is based on two important elements for measuring competitiveness; the competency of the companies and the quality of their strategies on the one hand (abilities of the private sector) and the extent of the quality of the business atmosphere nationally on the other.

Table 4.17 Kuwait's competitive positionTable 4.18 Most prominent strengths and weakness of the Kuwaiti Economy in accordance with the index of global competition 2006-2009

Globe Competition Index 2006-2007			Business Competition Index 2006-2007		
State	Table of Global Competition Report	Table of Sample of National Kuwaiti Competitiveness Commission	State	Table of Global Competition Report	Table of Sample of National Kuwaiti Competitiveness Commission
Singapore	5	1	Singapore	11	1
Norway	12	2	Norway	14	2
Estonia	25	3	Estonia	24	3
Emirates	32	4	Emirates	31	4
Slovenia	33	5	Slovenia	34	5
Slovakia	37	6	Slovakia	36	6
Qatar	38	7	Qatar	40	7
Kuwait	44	8	Kuwait	44	8
Cyprus	46	9	Cyprus	45	9
Bahrain	49	10	Bahrain	51	10

The State's general competition index reflects the quality level of a group of basic fundamentals for competition; a stable, economic framework, effective private and public institutions, suitable infrastructure, an educated workforce enjoying good health and opportunities for training, the competency of the markets and the ability to exploit available technological resources. In accordance with this index, Kuwait's position amongst the states listed in the world-wide report for competition is also 44 (Table 4.17) and Kuwait also occupies the eighth position amongst the sample States taken by the National Kuwaiti Competitiveness Commission after Qatar and the Emirates.

Kuwait's Competitiveness Commission also specified the strengths and weaknesses of the Kuwaiti economy based on a number of secondary indicators. We can summarize the country's strengths vis-à-vis global competitiveness in the following table (Table 4.18).

Table 4.18 Most prominent strengths and weaknesses of the Kuwaiti economy in accordance with the Index of Global Competition 2006-2007

Strengths		**Weakness**	
Element	International Position	Element	International Position
Budget Surplus	1	Restrictions on Foreign Ownership	125
National Savings Average	1	Direct Foreign Investment and Technology Transfer	119
Extent and Effect of Taxes	4	Innovative Ability	110
Brain Drain	8	Dependence on Professional Management	102
Flexibility in Determining Salaries	9	Government purchase of technological products	100
Government Debt	10	No. of Procedures for starting a commercial operation	94
Organized Crime	12	Effectiveness of the Boardrooms	93

On the basis of these indicators the report of the Kuwaiti National Competitiveness Commission calls on the government to upgrade the proficiency of the economy, improve the infrastructure (through investment in physical capital), restructure the educational system (to improve the quality of the human resource), expand the academic and technical infrastructure (to strengthen the information resource, improve the business environment to strengthen the private sector and reinforce the business ethic) and its associated legislation (Table 4.19).

Table 4.19 Most important general Economic Policies resulting from the Kuwaiti competitiveness report

General Economic Policies	Proposed Mechanisms and Initiatives
Upgrading the Proficiency of the Economy	1. Limiting Bureaucracy 2. More Decentralization in Decision Making 3. Establishment of a Competent Tax System 4. Guaranteeing a higher degree of Accountability in private and public institutions 5. Guiding the use of public funds to reduce levels of waste 6. Better co-operation
Improving the Infrastructure by Investing in Physical Capital Assets	1. Suitable programmes to raise the level of services so that they comply with highest standards 2. Improve port services 3. Improve communications network 4. Expand the water and electricity networks
Restructuring the Educational system to Improve the Quality of the Human Capital Resource	1. Import the quality of education 2. Raise percentages of primary school enrolment 3. Improve the internal competency of the educational system 4. Improve the quality of teaching staff 5. Improve the external competency of the educational system 6. Increase the role of the private sector in education and training

Expanding the Academic and Technical Infrastructure to strengthen the information resource	1. Increase National expenditure on research and development 2. Strengthen the relations between industry and academic research sectors 3. Mechanisms to increase innovative ability 4. Establishment of a complex for national excellence and talent
Improving the Business Environment to strengthen the Private Sector	1. Widen the base of economic diversification 2. Greater dependence on market systems and participation of the private sector 3. Development of the financial market 4. Increase the levels of competency in business administration 5. Guarantee greater protection of intellectual property rights 6. Issue anti-monopoly laws
Strengthening the Business Ethic and its Legislation	1. Guarantee greater transparency 2. Pass anti-corruption legislation 3. Adopt a special anti-corruption commission 4. Establish a special court for this purpose

4.18 The Kuwaiti Workforce

Recorded growth results indicate that the period 2002-2006 saw important developments at the level of the workforce. The total workforce in 2006 reached more than 1.87 million and the number of Kuwaitis in work rose to 6.7 annually on average from 258,540 employees in 2002 to 335, 238 in 2006. Women represented 51.4% of the total workforce compared with a percentage of 36.9% in 2002. The first indications showed that the formal education programmes (chapter 3) were bearing fruit. By 2002, the number of Kuwaitis holding higher and intermediate qualifications within the workforce had reached 59%. However, the overall job market is still suffering from the (continuing) concentration of Kuwaiti employees in the government sector, thereby escalating the scale of hidden unemployment. 85% of the total Kuwaiti workforces are employed in the government sector, whilst 15% approximately have made their way into the private sector. The focus and concentration of the expatriate workforce has continued to be in both the private and domestic sectors. The policy of bringing in expatriate workers shows that less than 19% hold higher and intermediate certificates, and there has been an increase in the percentage of expatriate workers without qualifications (to around 59% of the total workforce in 2006).

The increase of an uneducated expatriate workforce puts great pressure on public services subsidized by the State, and is considered as one of the challenges which the plan will have to deal with - and is due to employers having freedom to bring in expatriate workers.

The phenomenon represents a threat to growth and deep fragmentation has emerged in the Kuwaiti job market between increasing the employment of Kuwaitis in the private sector on the one hand, and achieving minimum costs of employment for the private sector on the other.

In recent years the government has worked on gradually rectifying the situation by implementing a number of reforms, the most prominent of which was the workforce reconstruction plan for the country. However, these reforms are of limited value, focusing as they do on the supply element of the national job market, and fail to move employment away from the traditional, government jobs. The government is determined to bolster reform by way of executive projects aimed at expanding privatization, freeing up the markets and developing the private sector to increase its ability to create more job opportunities. Also, the government aims to encourage the national workforce to take up employment in the private sector, by continuing to compel this sector to employ an increasing percentage of the national workforce.

Table 4.20 Distribution of the workforce by Academic Qualifications 2002-2006

	Kuwaiti Workforce			**Non-Kuwaiti Workforce**		
Year	Intermediate or less	Secondary	University or Above	Intermediate	Secondary	University
2002	41%	19%	40%	77%	12%	11%
2003	40%	19%	41%	79%	12%	10%
2004	39%	20%	41%	80%	10%	9%
2005	39%	21%	40%	82%	10%	9%
2006	40%	21%	39%	81%	10%	8%

Figure 4.13 The workforce by Academic Qualifications 2006

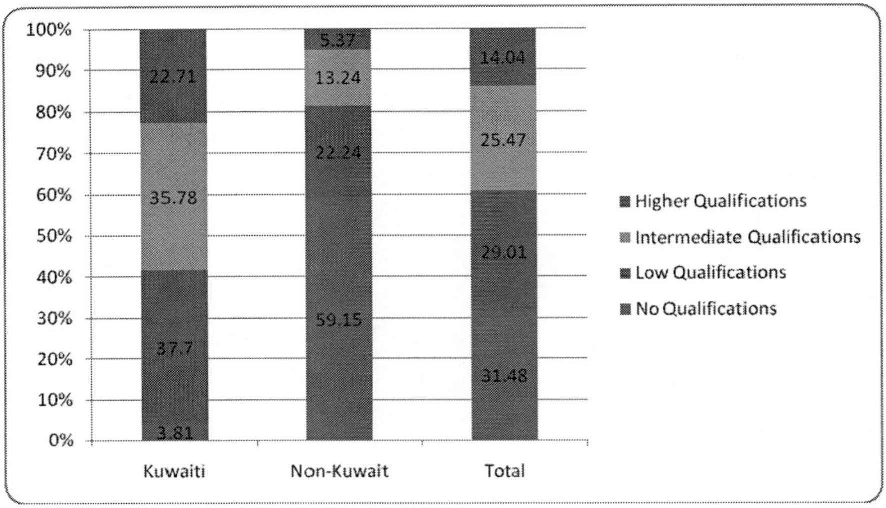

4.19 Innovation and Education

The human growth and development indicator prepared by the UN Development Programme has shown that Kuwait falls within the category of nations with high levels of human development, and the latest reports show that Kuwait clearly exceeds that of its counterparts amongst Arab countries.

Table 4.21 Educational characteristics of the people

Kuwaiti: K *Non-Kuwaiti: N-K*

Educational Level	**1985**		**1995**		**2006**	
	K	N-K	K	N-K	K	N-K
Above Secondary but below University	4.7%	2.4%	5.3%	2.7%	16.0%	10.2%
University and Above	5.7%	8.1%	9.3%	9.7%	7.9%	2.7%
Total:	**10.4%**	**10.5%**	**14.6%**	**12.4%**	**23.9%**	**12.9%**

Although the State has spared no effort in laying the groundwork for the success of the educational sector, as is evident from the rise in the student's average share of public expenditure on education, yet this sector continues to face a number of challenges especially at the level of internal proficiency.

The Kuwait report on competition made clear that one of the weak in the fields of Math's and applied sciences. The reason for this basically, is that the most important educational expenditure allocations are directed at the twin sectors of technical and administration. Only a small proportion of expenditure is allocated to education development research – methods of teaching, curriculum and teacher development - and these aspects are amongst the most important in the education system.

4.20 Human growth and development: health initiatives

As for the health sector, Kuwait during recent years has strengthened its relatively outstanding position amongst other states in the area of health service standards. Kuwait offers a basic system of health care which is considered by the standards of other Arab states to be advanced and the equivalent of what is offered in a number of advanced industrial countries. As is made clear by Table 4.22, the period 2002-2006 witnessed more important developments on a number of indicators connected with public health and health services.

Table 4.22 Development of Public Health and Health service indicators from 2002-2006

Public Health Indicator	2002	2003	2004	2005	2006
Expected life Span at birth of the citizens	74.95	75.36	75.02	76.87	78.02
Children under 5 death rates for every 1000 births	11.11	9.87	10.60	9.70	10.10
Infant death rates for every 1000 births	9.45	8.15	8.70	8.03	8.33
Death rates of mothers for every 1000 births	7.1	3.5	10.1	3.2	3.2
Percentage of births supervised by skilled health professionals	100%	100%	100%	100%	100%
Percentage of inhabitants who receive basic health care (improved health facilities)	100%	100%	100%	100%	100%
Doctors for every 1000 patients	1.68	1.74	1.78	1.86	1.94
Dentists for every 1000 patients	0.27	0.35	0.37	0.4	0.434

The public health indicators highlight significant developments the most important of which is that, amongst the Arabs, Kuwaitis have a longer life span. It is clear that the average anticipated life span at birth was around 78 years in 2006. This average mirrors that of advanced countries it is also clear there has been a significant drop in deaths amongst children under five, amongst suckling children and mothers. The health service indicators confirm that a 100% percentage has been maintained vis-à-vis the birth indicators and particularly those births supervised by skillful health operators. The 100 % figure is maintained in the indicator of citizens who have received primary health care. With regards to the indicator of the number of doctors per 1000 individuals, which is considered one of the most important indicators of health service proficiency, the indicator value rose from 1.68 doctors for every 1000 persons in 2002 to 1.94 doctors per 1000 individuals in 2006. This signifies that the numbers of doctors are on the increase and that there is a gradual improvement in health service. It is worth noting that the indicator value in this case is close to the international indicators in countries where health services are of the highest quality. The indicator value for dentists per 1000 individuals in Kuwait increased from 0.27 in 2002 to 0.434 in 2006. The Ministry of Health suffers from a lack of space and a deficiency in the number of foreign clinics. This situation increases the pressure on current health resources as well as the burden on primary health care and emergency services and it is hoped to remove these difficulties in the future.

In the field of women, women's share of employment opportunities at various levels witnessed some progress. The entrusting of a Kuwaiti woman with a ministerial position for the first time in the history of the country in 2005 was the most important success story in this field. However, the percentage of women represented in leading appointments below ministerial has remained low and steady at around 5% which does not reflect the percentage of women within the government apparatus nor the percentage of those holding high qualifications.

As far as co-operatives are concerned, their number reached 49 in 2006. The number of Kuwaiti citizens participating as members in these co-operations rose to 362,061 in the same year which is roughly 35.9% of the population. This expansion of the membership base enabled the co-ops to undertake various roles and to show a high degree of flexibility in facing up to the economic changes. In the area of civil society organizations, and as is evident in the following table, the period in question witnessed a tangible improvement especially in the total number of private societies (non-governmental) from 52 in 2002 to 79 in 2006.

This widespread concern reflects the importance attached to organizing the efforts of individuals in Kuwaiti society in the service of public issues including that of growth and development.

Table 4.23 Development of the civil society organizations from 2002-2006

Data	2002	2003	2004	2005	2006
Total No. of Private Societies (non-governmental)	52	52	53	75	79
No. of Charities	-	5	5	9	9
Societies Concerned with Specific Social groups:					
Children	3	4	4	4	6
Women	5	5	5	5	5
Specific fields (professional Societies)	15	15	15	15	15
Public Societies for the Preservation of the Environment	1	1	1	1	1
Professional Unions (Businessmen's Unions)	30	34	36	36	38

Despite some positive developments on the social level, several phenomena have emerged which are threatening the foundations and stability of Kuwaiti society and which the government trying to deal with is resolved to deal with. Amongst the most important phenomena of social challenges is the issue of increasing corruption as a socially accepted behavior and its negative impact on the structure of social values. Some indicators also point to the increasing phenomenon of breakdown in the role of the family and the resultant inconsistency in the structure of social values and to an intensity in the struggle between modern values and social inheritance. Issues of identity and problems of social violence have also emerged. The government, through proposals for legislation, mechanisms and programmes, will strive to help in the following areas:

To encourage social cohesiveness, activate the role of the individual, provide the opportunity for individual initiatives and strengthen social units and especially that of the family;

To increase the effective action of civil society organizations and the concern for issues of childhood, women, those with special needs and the elderly;

To ascertain issues of youth concern and their social problems and dealing with issues of leisure time; stimulate voluntary work, strengthen the value of participation and spread awareness of the relationship between citizen and state; and inculcate sound, social values and principles through education.

4.21 Conclusion

This chapter initially considered innovation and the diffusion of innovation within Kuwait. Since innovation depends on the strength of a strong demand from consumers and citizens for newer and creative ideas and innovative products and services, creating the optimal framework and possibilities to innovate becomes paramount. In essence there must be an acceptance for an innovation friendly market. In this regard, Kuwait is in need of new commitments. Economic metrics dictate the innovation initiatives. Economic measures that preserve established structures and organizations can diminish innovation initiatives. Consequently, countries like Kuwait could render themselves incapable of coping with the economic challenges that they face in a global economy if innovation is not nurtured and supported. Economic and growth metrics were provided to present Kuwait as an economy ready for the innovation take-off. The various sectors of Kuwait must be ready to invest in structural changes as Kuwait can only become comprehensively innovative if all sectors become involved. While this broad strategy needs to engage all parties – businesses, the public sector, producers and consumers – specific innovation-enhancing initiatives need to be focused on certain sectors. It also involves "buy-in" from key stakeholders. In the ensuing chapters, we examine the perspectives of these stakeholders on current innovation processes.

CHAPTER FIVE

CONCLUSIONS

"Innovation is the specific instrument of entrepreneurship. The act that endows resources with a new capacity to create wealth."
Peter F. Drucker

5.1 Introduction

Productive activity have always characterized the lives of the citizens of the Gulf and prior to the discovery of oil in particular. Innovation is a result of productive activity. During the periods of difficult and trying circumstances Gulf Countries tended to innovate and ensured the sustainability of human existence. Innovation could have been extensive during this period. This could have changed subsequent to the discovery of oil. The exploitation of natural assets is a matter of grave concern throughout the world, exploration and exploitation are costly and risky exercises in terms of growth and profitability and the risks are manifest in terms of social, political and economic consequences. Governments of oil rich Gulf Countries need to cultivate a culture that fosters creative ideas associated with, among others, safety and security of their natural and human resources, morality, employment and health within the context of an increasingly global environment. A lack of a shared vision, purpose and strategy reduces the vital role that innovation can play. A planned investment in innovation is therefore critical and Kuwait needs to reorient itself economically, politically, socially, ethically and morally in this regard. Failure to achieve this would result in Kuwait failing to fulfill its mandate, thereby reducing the return to private effort and initiatives.

This final chapter therefore presents the conclusions drawn from the literature survey and the empirical study. The idea that a 'decent' investment in innovation should become the true test of civilization is supported in this study. The chapter concludes with the major challenges that were encountered throughout the study and an outline for further research is provided.

5.2 Evaluation of innovation in Kuwait

The idea that innovation is vitally important for Kuwait's economic growth and human development has already been recognized throughout this study. The empirical findings support the general literature review that countries abundant in natural resources face the dilemma of how to effectively and productively manage sources of revenues accruing from the exploitation of natural resources. The literature and empirical findings suggest that resource abundance can cause weaknesses in the social, economic and (particularly) political environments where weak governance structures may result. The inability of Kuwait to grow and develop in accordance with classical economic growth theories where the national income is divided among labourers, landlords and capitalists in the form of wages, rent, interest and profits, suggests that an abundance of oil may result in curse effects. Classical economic theory raises the question of how an economy could be organized around a system in which each person sought his or her own (monetary) gain. Moreover, classical economic theories moved away from an analysis of the ruler's personal interests to broader national interests. Adam Smith in the Wealth of Nations saw national income as that produced by labour, land, and capital with property rights to land and capital held by individuals.

Although this should be a cause of great concern, there are many initiatives adopted by the current Kuwaiti government that reduce the effects of any possible curse. However, optimal innovative solutions need to be addressed. The purpose of this study was to present innovation as a curse reductive mechanism.

Kuwait derives almost all its budget directly from oil production, thus making Kuwait a rentier state. Being a rentier state also manifests itself to a certain extent in an adversity to democracy, with bureaucratic state machinery seeking to regulate markets. Rentier states can also oversee a poor innovation strategy. However, our empirical findings indicate that innovation in the finance and service sectors of Kuwait is more pronounced than in the oil sector and in this regard, innovation in the banking sector is identified as a critical driver for future growth and development. That the banking sector in Kuwait understands the role of innovation augers well for Kuwait in any policy making initiatives in other sectors.

One of the aims of the study was to analyze the importance of innovation, then determine which sectors innovate the best (why and how). Competition is propelling both corporations and governments to find new and alternative ways to increase productivity, and this is creating renewed interest in the need to innovate. The message for governments is that there is no substitute for good education and for policies that encourage investment in innovation, research and development. Innovation has a positive and beneficial effect on the performances of corporations and governments. However, a poor economic performance due to macroeconomic mismanagement by the government will result in investors losing confidence and this is something Kuwait needs to avoid. Table 5.1 shows the five best and worst ranked sectors in terms of innovation.

Table 5.1 Five best and worst ranked sectors in terms of innovation

Best		Worst	
1	Banking	1	Public Education
2	Private Communication	2	Public Health
3	Oil refinery and Petrochemical	3	Public Transportation
4	Finance, Insurance Real Estate and Business Services	4	Agriculture
5	Building and Construction	5	Public Communication

Conversely, the successes achieved by the five best sectors provide overarching framework for innovation. In support of the Government's objectives to transform and develop the economy, the banking sector has made substantial commitments to develop downstream industries. The provision of access to loan funding information and financing options are available through the banking sector. To build an innovative and sustainable future is the objective of public sector. The quest to transform the foundations of the Kuwait economy is the government's biggest contribution. However, public sectors featured more prominently as worst ranked sectors in terms of innovation. Yet within these sectors, private sectors have found business opportunities and begun to innovate (consideration to these specific initiatives adopted by the private sector remains outside the scope of this thesis).

5.2.1 Planning for outcomes relative to innovation

The empirical findings show that expenditure on research and development and investment in ICT are the important reasons why respondents selected the best performing sectors in terms of innovation. In any endeavor, planning for outcomes is critical. With specific reference to innovation, creating awareness among culturally sensitive stakeholders becomes an important element in the diffusion of innovation. Outcomes must be considered, and innovation must be based on an incremental approach involving a viable portfolio of projects. The implementation of a model for the management of innovation is critical if innovation is to succeed - especially in Kuwait. The empirical findings suggest that there is a strong belief that innovating improvement can now be achieved through foreign investment, whereas in the past the success of the Gulf States as identified by the Arab Planning Institute was seen as resting upon the large inflow of foreign labour. Innovation is uniquely human and is not random. One of the hallmarks of innovation systems is that it allows for the recruitment and retention of highly skilled human resources and, by allowing them access to knowledge, encourages them to discover new solutions to problems related to growth opportunities. In this regard a detailed plan for innovation needs to be considered by senior management to oversee the resource-allocation process. In their seminal work, Anthony, Johnson and Sinfield (2008) contend that it is companies that create blueprints for growth, construct innovation engines and support the engines with the right systems and mind-sets. Hence, planning for innovation requires institutionalizing innovation, and if Kuwait can build and maintain a capacity for innovation, substantial stakeholder benefits will accrue.

5.2.2 Assess, enable and transform current practices

One of the hallmarks of business is the separation of ownership and control. In Kuwait, ownership and control manifests in a distinct manner. Due to the philanthropic conduct of successive leaders, the government owns and controls substantial business interests, and following the failure of any major businesses, the government steps in with stimulus or rescue packages. Building on this unique approach towards creating a supportive infrastructure to promote an enterprising culture the government needs to assess the current practices. Developing a locally specific innovation model would undoubtedly need an assessment of the current position. Innovation does not take place in isolation. In the case of Kuwait this involves understanding the political and socio-economic environment. (This was identified in Chapter Three). There needs to be a buy-in by the Kuwaiti leadership, to first consider current practices and thereafter support stakeholder representatives to enable a culture of innovation to be nurtured. Ultimately, a transformation needs to take place. Kuwait's prolonged dependence on a single primary product export means that the labour-intensive stage of competitive industrialization must be leapfrogged. The empirical findings suggest that business more than government is an important agent of innovation. At the same time, the Kuwaiti Government is explicitly required to contribute to wider national development objectives, and this should not be forgotten. However, as there is a wide variation in how businesses and different industry sectors with diverse contexts and imperatives, interpret, approach and apply an innovation strategy.

Enhancing industrial transformation, as defined by International Human Dimensions Programme on Global Environmental Change (IHDP), becomes paramount in Kuwait. Industrial Transformation research commences with the notion that innovative changes in technology usage, change the ways in which humans use environmental resources and services. This embraces processes and products, production and consumption chains and distribution and disposal activities. In this regard the challenging goals of creating an awareness on how Kuwait should combine economic, political and social developments with the resource curse reduction, must be addressed. The scientific agenda should ideally address the following: alternative sources of energy and material flows; food security; water and transportation; information and communication; governance and political, economic and social transformation processes.

Kuwait therefore has to establish the critical issues that undermine the rate of adoption of technologies. More energy and materials intensive technologies should be avoided. Through consensus among the major stakeholders in the various different sectors, innovation can be advanced. Development of small, medium and micro enterprises would be crucial for an overall innovation strategy. A commitment to transform becomes important. A government fund could act as a business incubator and developer, yet successful business incubation in Kuwait requires enormous government and private sector support. Many time honoured traditions of the Kuwait government have played a major role in this regard and foreign investors are offered a transparent environment that guarantees the security of their long-term investments.

5.3 The revealing of dilemma facing Kuwait

The various chapters of this study have already identified that economies heavily dependent on natural resource exports have performed poorly on various measures of economic, social, and political development. This underscores the question of whether natural resources are a "curse" or a "blessing". The empirical evidence cited in this study suggests either outcome is possible. The curse effects included an appreciation of the real exchange rate, de-industrialization and limited growth prospects. The literature concurs that these adverse effects are more severe in volatile countries with weak institutions, corruption, autocratic governance, and underdeveloped financial systems. Fortunately, Kuwait has already been recognized as a highly ranked country in terms of the transparency international corruption perceptions index.

However, a resource boom reinforces rent grabbing and civil conflict, especially if institutional governance is weak. One of the challenges that face resource rich developing economies is the need to successfully move from depleting resources into developing other productive assets. The resource curse is therefore not inevitable, and good governance and sound economic policies can mitigate its ill effects.

In planning for economic growth, infrastructure development and an expansion of innovation should be a key priority. Unfortunately, any country endowed with rich natural resources will discover negative and positive impacts of certain developmental initiatives affect innovation. However, whether Kuwait favors an evolutionary process of innovation over the "leapfrog' process is yet to be established as while such 'leap-frogging' has been favored by some neighboring Gulf countries and in some cases this has led to indebted economies.

Should oil remain the energy source of choice for power generation and transportation, oil producing countries should recognize the need to maximize the wealth and longevity of their critical energy assets through effective management. The need to build infrastructures that will create a sustainable base for efficient exploitation of natural resources and concurrently allow for the development of a diversified economy will be imperative to ensure long-term prosperity. Investment in innovation therefore becomes paramount.

5.4 The need for innovation, diversification and knowledge creation

Kuwait's historic economic success cannot be sustained with the current levels of innovation. Innovation predicates a knowledge environment. If one considers current trends, the future of oil production in Kuwait appears firm and enduring as a source of energy and an important generator of revenues. Kuwait however increasingly needs to recognize the importance of maximizing the wealth and sustaining its critical energy assets through effective governance. The opportunity to invest wealth derived from natural resources in building a sustainable, efficient, diversified and innovative economy will enable Kuwait to make the most of its oil related assets far into the future.

Diversification helps support the profitable and sustainable development of oil and gas resources and long-run economic growth. Once diversification has commenced, it tends to gain momentum in a self-reinforcing cycle. A typical example would be Dubai that moved resources from oil production to develop a strong tourism related base. Consequently, several housing projects emerged for an international clientele. Moreover, the ports in Dubai have become a major hub for the movement of goods around the world. As one industrial sector's successes attract suppliers and skilled people, other sectors begin to locate nearby to take advantage of growing opportunities. Knowledge spills over from one sector to other sectors through individuals who migrate from one sector to another.

The challenges that Kuwait faces are somewhat complex. Therefore a fundamentally unique approach is required. Innovation and diversification takes place more and more beyond the boundaries of the country. In this regard new methods and tools are necessary to manage distributed knowledge creation. A cultural fit needs to be established.

5.5 Leveraging innovation for long term security

This study considered whether a country's natural resources are a curse or a blessing. The rate at which natural resources in Kuwait are exploited underline a cause for concern. From an economic perspective, how can a Kuwait endowed with rich natural resources switch from dependence upon these natural resources to develop other sectors based on knowledge, skills, capital and technology. Consequently, Kuwait could be called the "paradox of plenty" insofar as it is presenting unable to raise its growth rates and development levels above those of some countries lacking natural resources. Innovation may therefore be the key.

Innovation for Kuwait is non-negotiable. A consistent and predictable innovation process guarantees future growth and development. Consistency in innovation theory relates to changes in innovation being introduced gradually to accommodate stakeholders in various fields of endeavors and predictable relates to an unsurprising conduct on the beneficiaries of innovation. To achieve this five steps are proposed:

- Presenting a solid business case for innovation and creating a buy-in and an impetus for innovation that starts at the top;
- Reforming the attitudes of risk-averse individuals while promoting innovation in every department across the organization or state;

- Establishing a culture for innovation in order that a self-sustaining innovation process changes;
- Hiring innovators (as agents of change) to provide basic training in innovation; and
- Evaluating and measuring innovation results. Historically, incentives that encouraged innovation in the private sector of Kuwait tended to be less acceptable in the public sector where innovation generally lagged. This needs to change.

One important question relates to how innovation is understood and valued in cultural and social settings - and how different cultures can be motivated and supported to actively embrace innovation. While, in this regard, the three facets of intellectual capital (namely, human capital, social capital and cultural capital) need to be addressed, this issue falls beyond the bounds of the text. This text has examined how the core ideas associated with governance of scarce resources have been addressed within the context of the culture of the region. A large body of scholarship has a list of prescriptions for reducing the resource curse effects. The solutions suggested relate to macroeconomic issues, governance, economic diversification policies, natural resource funds, accountability and direct distribution as mechanisms for managing resource wealth effectively and efficiently. The study documents in a scholarly fashion the extent to which oil revenues accrue to Kuwait, and proposes that Kuwait has to become a better governor of these revenues. To ensure better governance, Kuwait has to enhance transparency, accountability, and become more supportive of a corporate culture, as a more robust institutional is needed framework to tackle potential 'curse' effects.

The philanthropic conduct of the rulers of Kuwait should in no way be belittled. The rulers have, over the years, shared the wealth of the country with their less fortunate neighbors. Expatriates living in Kuwait would readily attest to the conduct of the rulers in this regard. However, the question is whether such conduct can be sustained (and is so, how) is a cause for concern.

5.6 Implication for further work

Future research agenda on resource curse reduction through innovation for Kuwait needs to be firmly established. Researchers in Kuwait at institutions such as universities are ideally positioned to identify the multidisciplinary research needed to address further the issues raised in this study. However, one needs to move beyond mere philosophizing.

This research study has provided important insights into innovation as a reductive instrument for curse effects. However, the findings of this study suggest that while Kuwait can overcome the resource curse through innovation, political will among stakeholders is required. Privatization was an innovative solution to the resource curse in many countries. While it is a possible solution for combating the resource curse, this study favours developing a strategy to directly address the pervasive problem of weak institutions in Kuwait that does not favor innovation.

Finger pointing is not a way forward. Collectively, the private sector and the public sector must develop a strategy to innovate that will sustain the wealth of the country.

5.7 The challenges encountered

There were numerous challenges that were encountered during the period of the study. The study evolved as a result of a passion that I have for 'reinventing' Kuwait through a process of innovation. However, as these thoughts developed completion of a doctoral programme was considered appropriate.

Preparing work on the thesis introduced me to the notion of the resource curse. The resource curse has been identified in numerous studies written across the developed world. However, in these developing economies endowed with rich natural resources. There has been much less research on recognition of these effects. Strategic studies in the Gulf addressing the curse effects are not readily available and, to the best of my knowledge, this is the first study in Kuwait. It is an intention of the author to ensure that more research should be published in this area, as it is certain that this topic would generate further discussions and ideas within the country.

The methodology adopted for this study was outlined in chapter Five. Although initial discussions were held with key stakeholders, their views on innovation and profits are almost certainly determinant upon past practices – when rulers of Kuwait tended to overlook the need for profitability and innovation in favour of welfare provision based on an oil economy. Nevertheless it is important to confer with such key individuals given their current and future economic roles.

Questionnaires were provided in both Arabic and English. While translating the questionnaires and the responses proved time-consuming the major challenge encountered were associated with translating the findings of this research into more meaningful outcomes that would engender a culture of innovation.

5.8 Implications for future research – innovation in relation to productivity

Innovation is a key source of productivity, growth and development. Growth and development as reductive curse measures were adequately addressed throughout this study. However, the impact of innovation on productivity was not addressed. While this study focuses on the innovation dynamics in general, one should recognize that is not the sole determinant of economic growth and development. An approach that allows one to conceptualise innovation as a deliberate and specific process of change in a particular firm, needs to be considered in future studies. Following upon this one needs to revisit the role of the government and private sector in incubating small and medium sized enterprises.

Capital in the form of financial, natural and physical are the ones that are commonly factored into growth models. In any discussion relating to innovation in Gulf Countries, human, cultural and social capital should also be considered. Human capital is generally considered as the knowledge, skills and competences embodied in individuals that are relevant to economic activity. An investment in human capital can be an important determinant of innovation. An individual's (or groups of individuals) earning capacity and employment plays an important role in determining the level and distribution of income in society and, by extension, the pace of in economic growth. A degree of government and private involvement is needed in the provision of educational services. Ensuring that Kuwaitis enter their working life with a certain level of human capital acquired during the years of compulsory education is paramount. However, Kuwaitis are also heavily dependent upon the government's financing and delivery of post-compulsory education, where returns may - to a larger extent- accrue to the individual, but society also benefits. In Kuwait, an investment in education allows one to move from non-elite positions into elite positions. This is particularly necessary for the diffusion of innovation.

Secondly, cultural capital, on the other hand, is a more academic and philosophic notion, referring to the credentials and cultural assets embodied in individuals and their families. Cultural capital is a notion that explains disparities in the educational attainment of children from different social classes. Understanding cultural issues in Kuwait are fundamentally important to innovation success. Since culture shares many of the properties that are characteristic of economic capital, cultural capital - in the form of habits and customs - comprises a resource capable of either generating or inhibiting returns. As, cultural capital can be monopolized by individuals and or groups, and transmitted from one generation to the next the concept can be used to explain the reproduction of social hierarchy. Extending this notion of cultural capital further, countries that are well governed and show a high concern for human wellbeing also endow their citizens with positive cultural capital.

Thirdly, social capital defined in terms of faith, trust in other people, and prevailing social norms, allows agents and institutions to be more effective in achieving common objectives. The most common measures of social capital include one's participation in various forms of civic engagements, membership of voluntary associations, civil societies and political organisations. Social capital has been deployed to explain a wide range of social phenomena, including general economic performance, levels of crime and disorder, immigrant employment and health trends. Moreover, social capital can have socially undesirable effects, where trust and mutuality operate to enhance inequalities, exclusion or even criminality.

There is thus a need to take the innovation discussion further to accommodate issues associated with human capital, social capital and cultural capital. Investing in innovation should ideally entail a return to these forms of capital too. Needless to say that failure to invest in these forms of capital would have devastating consequences on innovation.

5.9 Common good for all

The leadership of Kuwait have at all times maintained a humanitarian approach towards it citizens. The political and constitutional imperatives of the country are set ensure that the citizens of the country enjoy the economic, political and social stability of the country. There are many challenges that the leadership faces in pursuing common good objectives. In the event individual interest supersedes those of national interest, situations could result in instability as we have seen in other countries associated with poor governance. Common good is a good proper to, and attainable only by the community, yet individually shared by its members. Oil in Kuwait therefore becomes a common good. Managing a common good today is becoming a major challenge. The changes in demography together with varying political, social and economic outlooks are considered major issues in this regard. To this end Kuwait has been able to maintain a certain sense of calm by their continued philanthropic conduct.

5.10 Final thoughts

The pursuit of political emancipation has long been the objectives of many colonized countries. Today economic emancipation, freedom from government intervention and a move away from a dependence syndrome are issues that government and the private sectors need to address. Innovation can play a critical role in these discussions. This final and concluding chapter presents the recommendations of the study. Having identified the role that innovation can play, the chapter advocates the further development of technologies and infrastructures that would enable Kuwait to exploit its mineral wealth efficiently -and at the same time provide a platform for economic diversification. Since management of the revenues of the oil industry are critical to the success of these efforts, a more robust and multi-disciplinary approach to innovation is recommended. Any such move requires collaboration among industry, government, and research/educational institutions, and the promotion of a culture of innovation and creativity. Consequently, communication among all sectors becomes crucial. Innovation will enable all sectors to exploit their potential. This expansive study is the very first one to have linked the resource curse to the innovation literature. Indeed it is the first study to have been conducted by a Kuwaiti national into the specificities of the natural resource of the country and innovation at the sectorial level. The challenge is now to take these ideas forward.

BIBLIOGRAPHY

Abernathy and Utterback (1978)'Patterns of Industrial Innovation', in Technology Review 80(7), pp. 40-47.

Abu-Hakima, A.M (1983) 'The Modern History of Kuwait', Luzac & Co. London.

Adams et. al., (2006) Innovation Management Measurement: a Review, International journal of Management Reviews, Vol.8.

Afsaruddin, A. (2002) Excellence and Precedence: Medieval Islamic Discourse on Legitimate Leadership. Leiden: E. J. Brill.

Afuah, A. (2003) Innovation Management: Strategies, Implementation and Profits. Oxford University Press Inc.

Al-Hassan, A., Delgado, F. and Omran, M. (2007) IPO Behaviour in GCC Countries: Goody-Two Shoes or Bad-to-the-Bone? IMF Working Paper, Washington DC: IMF.

Al-Hijji, Y.Y. (1997) Old Kuwait: Memories in Pictures. Almansouria, Kuwait: Center for Research and Studies on Kuwait.

Al-Melhem, A. (1998) 'Privatisation and Protection of Investors in Kuwait: Reality and Ambition', Arab Law Quarterly, 13(2), pp. 178–99.

Al-Mughni, H. (2001) Women in Kuwait: The Politics of Gender. London: Saqi Books.

Al-Sabah, M. (1988) 'The Dutch Disease in an Oil Exporting Country: Kuwait', OPEC Review (Summer), pp. 129-144.

Anthony, S.D. et al. (2008) The Innovator's Guide to Growth: Putting Disruptive Innovation to Work. United States: Harvard Business School Press.

Auty, R.M. (1994b) 'The resource curse thesis: minerals in Bolivian development, 1970-90', Singapore Journal of Tropical Geography, 15(2), pp. 95-111.

—. (1998) Resource Abundance and Economic Development: Improving the Performance of Resource Rich Countries. Helsinki: The United Nations University/World Institute for Development Economics Research.

—. (2001) 'The Political Economy of Resource-Driven Growth', European Economic Review, 45, pp. 839-946.

—. (2001b) 'Why resource endowments can undermine economic development: concepts and case studies'. Paper prepared for the BP-Amoco Seminar. Lincoln College Oxford University. November 28-29.

—. (ed.) (2001a) Resource Abundance and Economic Development. Oxford: Oxford University Press.

Auty, R.M. and Gelb, H.A. (2001) 'The Political Economy of Resource-Abundant States', Resource Abundance and Economic Development, Oxford University Press, pp. 126-144.

Auty, R.M. and Mikesell, R.F. (1998) Sustainable Development in Mineral Economies. Oxford: Clarendon Press.

Baaklini A. I. (1982) Legislatures in the Gulf: The experience of Kuwait, 1961 –1976, International Journal of Middle East Studies, vol. 14 No. 3, Cambridge University Press. Pages 359-379.

Bakken, T. and Hernes, T., 2002, 'Introduction: Niklas Luhman's autopoietic theory and organization studies – a space of connections' Oslo, 9-22.

Berkhout, A.J. et al. (2006) 'Innovating the innovation process', International Journal of Technology Management, 34(3/4), pp. 390-404.

Berry, M. M. J. and Taggart, J. H. (1994), Managing Technology and Innovation: a review. R&D Management, 24: 341–353.

Bessant et al. (2000)"New elements in national /regional innovation systems to support SMEs - Developing learning networks, in Innovation and Regional Development, M. Heitor, Editor. 2000, ITEC: Lisbon.

Blair, T. (2010) Vision Kuwait 2035 Report: Executive Summary. Kuwait.

Blas, J. and Kerr, S. (2007) 'Gulf Consumer Spending Soars Despite Weak Dollar', Financial Times, 24 July.

Bondarevsky, G. (2003) *'Mubarak's Kuwait in Russian and German Policy'*, in Slot B.J. (2003*)Kuwait: The Growth of a Historic Identity*. London: Arabian Publishing Ltd, pp. 49–57.

Boschini, A.D., Pettersson, J. and Roine, J. (2005) Resource curse or not: A question of appropriability. http://unixware.mscc.huji.ac.il/~melchior/DEGIT/69_Boschini_et_al_Resource_Curse.pdf

Burns. T., and Stalker, G.M, (1961), The Management of Innovation. London : Tavistock.

Carneiro, F.G. (2006) Development Challenges of Resource-Rich Countries: The Case of Oil Exporters. Proceedings of the VI

International Colloquium.Working Paper Series.The World Bank, pp. 115-140.
Carter C. F. and Williams B. R., (1959) The Characteristricts of Technically Progressive Firms, Journal of Industrial Economics, 7(2), pp. 87-104.
Castells (1992) Educate~ Vol.6, No.1, 2006, pp. 6-12
Chesbrough, H. (2003), Open Innovation: the new imperative for creating and profiting from technology, Harvard Business School Press, Boston, MA.
Chesbrough, H.W., (2003c). The era of open innovation. MIT Sloan Management Review 44 (3), 35–41.
Chesbrough, H.W., Crowther, A.K., (2006). Beyond high-tech: early adopters of Open Innovation in other industries. R&D Management 36 (3), 229–236.
Chesbrough, Henry W. (2003b) The Era of Open Innovation. Sloan Management Review.
Christensen Clayton M., and Michael E. Raynor (2003)" The Innovator's Solution", Creating and Sustaining Successful Growth.
Cohen. W.M. and Levinthal, D.A. (1989) 'Innovation and learning: The two faces of R&D', The Economic Journal, 99, pp. 569-596.
Collier, P. and Hoeffler, A. (2000) Greed and Grievance in Civil War. Policy Research Working Paper 2355, Development Research Group. Washington DC: The World Bank.
Corden.W.M. and Neary, J.P. (1982) 'Booming Sector and De-industrialisation in a small open economy', The Economic Journal, 92, pp. 825-848.
Cowen, T. (2007), A way for Resource-Rich Countries to Audit Their Way Out of Corruption.
Coxhead, I. (2006) A New Resource Curse? Impacts of China's Boom on Comparative Advantage and Resource Dependence in Southeast Asia, World Development Vol. 35, No. 7, pp. 1099–1119.
Crystal, J. (1995) Oil and Politics in the Gulf: Rulers and Merchants in Kuwait and Qatar. Cambridge, UK: Cambridge University Press.
Cyert R.M., and J.G. March, (1963/1992) A Behavioral Theory of the Firm.2nd ed. Prentice Hall, Englewood Cliffs, NJ.
David, P. and Forey, D. (1995) 'Assessing and Expanding the Science and Technology Knowledge Base', Science and Technology Industry Review, 16, OECD, Paris.
De Boer, K. and Turner, J.M. (2007) 'Beyond oil: Reappraising the Gulf States', The McKinsey Quarterly, 31 January, pp. 7-17.

De Ferranti, D. et al. (2002) From Natural Resources to the Knowledge Economy: Trade and Job Quality. Washington DC: The World Bank.

Delgado, M., Porter, M. and Stern, S. (2008) Convergence, Clusters and Economic Performance.Working Paper.

De Soysa, I. (2002) Ecoviolence: Shrinking Pire or Honeypot? Global Environmental Politics, 2(4), pp. 1-36.

Dhanaraj, C. and Parkhe, A.(2006) 'Orchestrating innovation networks', Academy of Management Review, 31(3), pp. 659–669.

Dicken, P. (1998) Global Shift.3rd edition. New York: The Guilford Press.

Drucker, P. (1992) The age of discontinuity: Guidelines to our changing society. Harper and Row. New Jersey.

Domar E.D. (1946) "Capital Expansion, Rate of Growth, and Employment", Econometrica, 14, April, pp 137-47.

Engeli, G. and Pieth, M. (2000) Developing an integrity programme for FIDIC: a private sector initiative to prevent corruption in IFI-funded public procurement, Background Paper, Commissioned by the World Bank and prepared for the Annual Meeting of FIDIC 2000 in Honolulu.

Evans, P.B. (1989) 'Predatory, developmental, and other apparatuses: a comparative political economy perspective on the Third World state', Sociological Forum, Vol. 4, No. 4, 561-87.

Fagerberg, J., Srholec, M. and Verspagen B. (2009) *Innovation and Economic Development*, TIK Working Paper on Innovation Studies No. 20090723, Centre for technology, innovation and culture, Oslo.

Fields, G. (1989) 'Change in poverty and inequality in the developing countries', World Bank Research Observer, 4(2), pp.167-85.

Frankel, J. (2010) The Natural Resource Curse: A Survey, Faculty Research Working Paper Series, Harvard Kennedy School.

Freeman, C. (1982) The Economics of Industrial Innovation, 2nd edn, Frances Pinter, London.

Garaibeh, H. (1987) 'Government Income Sources and the Development of the Taxation System – The Case of Jordan, Egypt and Kuwait', in H. Beblawi and G. Luciani (eds), The Rentier State: Volume II, London: Croom Helm.

Garnaut, R. (2010), 'Climate Change, China Booms and Australia's Governance Struggle in a Changing World', 2010 Hamer Oration, Melbourne, 5 August;

Godin Benoit (2003), The New Economy: What the Concept Owes to the OECD, Project on the History and Sociology of STI Statistics, Paper no. 21, 24 pages. Published in Research Policy, 33 (2004): 679-690

Gupta, A.S., 2007. Determinants of tax revenue efforts in developing countries.International Monetary Fund Working Paper 07/184, Washington, DC.

Gylfason. T., Herbertson, T.T. and Zoega, G. (1999) 'A mixed blessing: natural resources and economic growth', Macroeconomic Dynamics, 3(2), pp. 204-225.

Harford, T. and Klein, M. (2005) Aid and the Resource Curse: How Can Aid Be Designed to Preserve Institutions? Public Policy for the Private Sector, April. Washington DC: The World Bank.

Harrod Roy, (1949) "Wesley Mitchell in Oxford".Economic Journal, 60 (September), 459-60.

Heuty, A. and Carlitz, R. (2008) Resource Dependence and Budget Transparency, Revenue Watch Institute, Janurary 2008

Jamal, M.A.H. (1998) History of Postal Service in Kuwait. Kuwait City: Center for Research and Studies on Kuwait.

Jennings, S.F., and Moore, S.A. (2000) The rhetoric behind regionalization in Australian natural resource management: Myth, reality and moving forward. *Journal of Environmental Policy and Planning* 2: 177-191.

Karl, T.L. (1997) The Paradox of Plenty: Oil Booms and Petro States. Berkeley, CA: University of California Press.

Karnik, A. and Fernandes, C. (2009) 'Natural resource dependence: a macroeconometric model for the United Arab Emirates', Applied Economics, 41(9), pp.1157-1174.

Katz, R., and von Hippel, E. (2002)."Shifting Innovation to Users Through Toolkits." Management Science, 48(7), 821-33.Oxford University Press.

Kelly, P. and Kranzberg, M. (1978) Technological innovation: A critical review of current knowledge. California: San Francisco Press.

Kline, S.J. and Rosenberg, N. (1986) The Positive Sum Strategy, Harnessing Technology for Economic Growth: An Overview of Innovation. Washington D.C. USA: National Academy Press.

Kondratieff "The Study of Innovations" Innovation Management and New Product Development.

Leith, J. C. (2005)Why Botswana Prospered? Montreal: McGill-Queen's University Press

Lewin, M., (2011)Botswana's Success: Good Governance, Good Policies, and Good Luck.
http://siteresources.worldbank.org/AFRICAEXT/Resources/Botswana_success.pdf

Limi, A., (2006) Did Botswana Escape from the Resource Curse? pp. 2/3 International Monetary Fund.WP/06/138. IMF Working Paper

Lundvall, B.A. (1992) National Systems of Innovation.Towards a Theory of Innovation and Interactive Learning. London: Pinter Publishers.

Mehlum, H., Moene, K. and Torvik, R. (2002) 'Plunder & Protection Inc.', Journal of Peace Research, 39(4), pp. 447–59.

Mikesell, R. (1997) 'Explaining the resource curse, with special reference to mineral-exporting countries', Resources Policy, 23(4), pp. 191-199.

Moshaver, Z. (2003) Revolution, Theocratic Leadership and Iran's Foreign Policy: Implications The Review of International Affairs, Vol.3, No.2, Winter Edition.

Mottahedeh, R.P. (1980) Loyalty and Leadership in an Early Islamic Society.Princeton, NJ: Princeton University Press.

Myers. S. and Marquis, D (1969), "Successful Industrial Innovations", NSF repot N. 69-17, National Science Foundation, Washington, DC.,

Nelson (1993) National Innovation Systems, Oxford University Press, Oxford.

Newman, H.R. (2003) The Mineral Industry of Spain, U.S. Geological Survey Minerals Yearbook, 2003.

Nonini, D. M. (2008) "Is China Becoming Neoliberal?" Critique of Anthropology 28(2): 145-176.

Palliam, R. (2006) "Further evidence on the information content of economic value added", Review of Accounting and Finance, Vol. 5 Issue: 3, pp.204 – 215.

—. (2010) Investment in Long Term Security Taken to Another Degree, Annual report of the Gulf Studies Center, American University of Kuwait.

Palliam, R, Cader, H, A and Chiemeke, C. (2011) Succession issues among family entrepreneurs in the countries of the Gulf, International Journal of Business Administration, Vol. 2, No. 2, May 2011.

Palliam, R., Galfy, A.H. and Palliam, S.R. (2010) Investigating risk-return profile of graduates: a corporate accountability and social responsibility perspective, Journal of Business Research, March, 2010.

Parkhe, A. et al. (2006) 'New frontiers in network theory development'. Academy of Management Review, 31(3), pp. 560-568.

Paya, A. (2010) Iraq, Democracy and the Future of the Muslim World, Taylor and Francis.

Van Der Ploeg, F. (2011) Natural Resources: Curse or Blessing?Journal of Economic Literature 49:2 pp. 366-420.

Porter M. and Stern S., (2008). Surveys and Archival Analysis Studies of national innovation systems by ... volume has been published on issues in multilevel analysis of creativity and innovation in business and beyond social science ---Page 251

Porter, M.E. (1990) The Competitive Advantage of Nations. Macmillan: London

—. (1999) 'The Economic Performance of Regions', Economics of Innovation and New Technology, 8, pp. 5–25.

Posnikoff, J. 1997. Disinvestment from South Africa: They did well by doing good, Contemporary Economic Policy, 15: 76-86.

Prebisch, R. (1950) The economic development of Latin America and its principal problems. Lake Success, N.Y.: United Nations.

Raghuram, R. G. and Subramanian, A. (2005) What Undermines Aid's Impact on Growth? Working Paper, No. 05/126, International Monetary Fund, Washington, DC.

Ross M. Stolzenberg 2006. *Sociological Methodology, 2006.* Oxford: Basil Blackwell for the American Sociological Association.

—. 2005. "Analytical Methods for Improving Common Survey Measures of Money Income" pages 510-513 in *Working Families: Time Apart, Time Together* edited by B. Schneider and L. Waite. Cambridge, England: Cambridge University Press.

Ross, M. (2001) Extractive Sectors and the Poor.An Oxfam America Report. [Online]. Available at: http://www.oxfamamerica.org/eirexport/index.html (Accessed: 1 January 2010).

Rothwell, R. (1992) 'Developments Towards the Fifth Generation Model of Innovation', Technology Analysis and Strategic Management, 1(4), pp. 73-75.

Rush, A. (1987) Al-Sabah: Genealogy and History of Kuwait's Ruling Family, 1752 – 1986. London and Atlantic Highlands N.J.: Ithaca Press.

Sachs, J.D. and Warner, A.M. (1995a) *'Economic convergence and economic policy'*.NBER Working Paper, No. 5039. Cambridge, MA: National Bureau of Economic Research.

Sachs, J.D. and Warner, A.M. (1995b) *'Natural resource abundance and economic growth'*.NBER Working Paper, No. 5398. Cambridge, MA: National Bureau of Economic Research.

Sachs, J.D. and Warner, A.M. (1997) 'Natural resource abundance and economic growth'.Center for International Development and Harvard Institute for International Development. Cambridge, MA: Harvard University.

Sachs, J. D. and Warner, A. M. (2001) 'The curse of natural resources', European Economic Review, 45, 827–838.

Salisu, S.M. and Yagudin, S.Y. (2007) Investments and Vertical Integration - Catalyst for Development in the Oil Industry of an

Emerging Market: The Case of Refining in Kuwait, Oil and Gas Business.
Sarraf, M. and Jiwanji, M. (2001) 'Beating the resource curse: the case of Botswana', Environmental Economics Series, Paper No. 83. Washington, DC: The World Bank.
Schumpeter J.A. (1934) The Theory of Economic Development. Cambridge, Mass.: Harvard University Press.
—. (1939) "Business Cycles defined economic evolution in a very broad fashion:"The changes in the economic process brought about by innovation together with all their effects, and the response to them by the economic system..." (p. 86).
—. (1942) Architectural Innovation: The reconfiguration of existing product technologies and the failure of established firms.
—. (1943) Capitalism, Socialism and Democracy. London: Allen and Unwin.
Simon H.A., Models of Man, New York: Wiley, 1957.
Sinfield Joseph V., (2008) " The Innovators Guide to Growth".
Singer, H. (1950) 'The Distribution of Gains between Investing and Borrowing Countries', The American Economic Review, 40(2), pp. 473–85.
Smith, A. (1776) An Inquiry into the Nature and Causes of the Wealth of Nations. USA: Oxford University Press.
Smith, B. (2001) Rethinking the politics of the resource curse: oil wealth and political stability.Unpublished mimeo.
Star Paul, "The Meaning of Privatization," Yale Law and Policy Review 6 (1988): 6-41
Stevens, P. (2003) Resource Impact: Curse or Blessing? A Literature Survey.
Taguchi H. and Nonaka I., 1995.The Knowledge-Creating Company. New York: Oxford University Press, 1995. (particularly pp.44ff.)
Trott, P. (1995), "Inward technology transfer as an interactive process: a case study of ICI", Technovation, Vol. 15 No.1, pp.25-43
—. (2005) Innovation Management and New Product Development.3rd edn.
United Nations Commission on Sustainable Development (2008) Economic Development, Economic and Social Council, Report on the Sixteenth Session, 20th & 21st Meetings. New York: United Nations.
Vandergeest, P., and Peluso, N. L., (2006) Empires of forestry: professional forestry and state power in Southeast Asia, Part 1. *Environment and History 12: 31–64*. Online, accessed 16 July 2011.

Van der Panne, G. and Van Beers, C. (2006) 'On the Marchall-Jacobs controversy: it takes two to tango', Industrial and Corporate Change, Oxford University Press, 15, pp. 877-890.
Van der Panne, G., Van Beers, C. and Kleinknecht, A. (2003) 'Success and failure of Innovation: a literature review', International Journal of Management, 7 (2), pp. 1-30.
Van Schalkwyk, M., (2008) Government Outlines vision, Strategic Direction and Framework for Climate Policy.Statement by Marthinus van Schalkwyk, Minister of Environmental Affairs and Tourism, 28 July, Cape Town. www.environment.gov.za, accessed 16 July 2011.
Vine, P. and Casey, P. (1992) Kuwait: A Nation's History. London: Immel Publishing Ltd.
Von Hippel, E. (2001) Perspective: User toolkits for innovation. Journal of Product Innovation Management, 18(4), pp. 247-257.
Wegenast, T. (2007) "Inclusive Institutions and civil war onset in resource-rich countries" Paper presented at the annual meeting of the Midwest Political Science Association 67th Annual National Conference, The Palmer House Hilton, Chicago, IL.
Weil, Gotshal, and Manges (2002) Swedish Shareholders Association, Corporate Governance Policy.
Woodward, J. (1965) Industrial Organization, London: Oxford University Press.
Woolgar et al.'s (1998) "The International Handbook on Innovation". Universities and Outside Research In framework, universities fall well outside of a small firm's focus of ... Evidencing the differences between small firms----Page 669
World Bank Archives, World Bank Group Historical Chronology. Available at: http://web.worldbank.org/WBSITE/EXTERNAL/EXTABOUTUS/EXTARCHIVES/0,,contentMDK:20035653~menuPK:56305~pagePK:36726~piPK:36092~theSitePK:29506,00.html (Accessed: 1 January, 2010).
Workman, T. (1991) The social origins of the Iran-Iraq war, York University.
Wright, G. and Czelutsa, J. (2002) Resource-based economic growth, past and present. Stanford, USA: Stanford University Press, pp. 34-37.
Yam, R.C.M., Lo, W., Tang, E.P.Y., and Lau, A.K.W. (2011) Analysis of sources of innovation, technological innovation capabilities, andperformance: An empirical study of Hong Kong manufacturing industries, Research Policy 40 (2011) 391–402, Elsevier.

Zahra, S.A. and George, G. (2002) 'The net-enabled business innovation cycle and the evolution of dynamic capabilities', Information Systems Research, 13, pp. 147-150.